STUDY GUIDE TO ACCOMPANY

MODERN PHYSICAL GEOGRAPHY

FOURTH EDITION

ALAN H. STRAHLER
ARTHUR N. STRAHLER

JOHN WILEY & SONS, INC.
New York • Chichester • Brisbane • Toronto • Singapore

ISBN 0-471-55103-1

Printed in the United States of America

10 9 8 7 6 5 4 3 2

Camera-ready copy prepared by Kristi S. Strahler

Printed and bound by Malloy Lithographing, Inc.

HOW TO USE YOUR STUDY GUIDE

WHAT YOUR STUDY GUIDE CONTAINS

Your Study Guide includes :

- A *Concise Summary* of all important facts and concepts from the textbook.
- *Definitions* of all important terms in the textbook.
- *Self-Testing Questions* on facts, concepts and terms.
- *Sample Objective Test Questions,* including matching, multiple-choice, and fill-in questions.

HOW TO USE YOUR STUDY GUIDE

Take the *Study Guide* to class with you and follow the lecture topics as they are presented. Underline terms, concepts, and facts stressed by your instructor. Make marginal notes of special items or examples. If the instructor is omits a topic, draw a line through the part of the *Study Guide* text.

Within a few hours after your class lecture, read the textbook chapter carefully. Place the *Study Guide* pages beside the book and compare the *Study Guide* items with the textbook. Refresh your memory of your instructor's lecture explanations. Note in the *Study Guide* any points that remain unclear, and bring these to your instructor or the teaching assistant at the next possible opportunity.

We emphasize that you should read the textbook carefully after each lecture, while things are still fresh in your mind. Don't let this reading pile up until just before a unit test or the final examination.

After reading the textbook chapter, use the *Study Guide* in the following way to check your knowledge. Using a sheet of blank paper or a piece of cardboard, cover the right-hand column of the page, exposing only the questions in the left-hand column. Read each question and recite the answer to yourself, preferably out loud. When you have given your best answer, pull down the covering sheet and expose the lines opposite the question. If you need more information or reference to a diagram, refer to the textbook page number given at the extreme right.

In using the *Study Guide's* self-testing questions, don't simply think to yourself, "Sure, I know all that. What's next?" You must make the effort to compose an answer in words, as if you were telling it to someone else. Speak the answer out loud if you can, or at least say the words silently to yourself. If your practice is to study with a companion, take turns reciting the answers. Your partner can check your answer against the *Study Guide* and prompt you on points omitted. When you make this effort, you will quickly discover that you have difficulty in recalling certain important words, or that you really aren't sure of their explanation. Study them again, using the *Study Guide* and textbook, until you have them correct.

SAMPLE OBJECTIVE TEST QUESTIONS

Each chapter in your *Study Guide* has a set of sample objective test questions. Use this test after you have completed your chapter study and oral self-testing program. Mark the answers in the blanks. Then compare with the correct answers given at the back of the book. Here are some tips: On the word-matching questions, there may be a term in the right column that applies to more than one term in the left column. However, there is only one best way to match all terms in the list to make the closest connections in meanings. In the multiple-choice questions, don't stop after marking the first answer that looks good. Read the remaining choices and reject them, so as to reinforce the correctness of your choice. Only one of the choices is entirely accurate, although there may be some measure of truth in others. The completion question puts more stress on you than other types, because it depends upon recall, not upon recognition. This type of question tells how well you have mastered the definition of important technical terms.

Keep Up With Your Study and Self-Testing!

Don't Fall Behind!

Don't Let Things Pile Up to the Very Last!

Get Help Quickly on any Topic You Don't Understand!

CONTENTS

INTRODUCTION

PAGE

PHYSICAL GEOGRAPHY 1

What is *physical geography?*

Physical geography: study and synthesis of the important elements of human physical environment.

What branches of natural science are included in physical geography?

Branches of natural science included in physical geography are atmospheric science (meteorology, climatology), physical oceanography, geology, soil science, plant ecology, biogeography, and geomorphology.

What aspect of these sciences is emphasized in physical geography?

Spatial relationships are emphasized, including the systematic arrangements of environmental elements into regions over the earth's surface and the causes of those global patterns.

THE LIFE LAYER 1

What is the *life layer?*

Life layer: shallow surface zone of the lands and oceans containing most of the world of organic life, or *biosphere.*

Define *biosphere.*

Biosphere: total of all forms of living organisms of our planet.

What is the *atmosphere?*

Atmosphere: gaseous envelope or layer surrounding the solid earth, in direct basal contact with the land and ocean surfaces.

In what ways is the atmosphere important to the human physical environment?

The atmosphere determines climate of the life layer and supplies vital life elements—carbon, hydrogen, oxygen and nitrogen—to the biosphere.

What is the *lithosphere?*

Lithosphere: general term for the entire solid earth (note: In geology, "lithosphere" refers only to the outermost rock layer of the earth; a brittle layer of which moving "lithospheric plates" are formed.)

How is the lithosphere important to the human physical environment?

The lithosphere forms a stable platform for the life layer. The surface of the lithosphere is shaped into landforms that provide varied habitats for plants. The rock of the lithosphere supplies many nutrient elements essential for plants and animals.

What is the *hydrosphere?*

Hydrosphere: collective term for all of the earth's free water in liquid, solid, and gaseous forms.

In what way is the hydrosphere essential to life on earth?

All living forms of plants and animals require water to sustain life.

Define physical geography in terms of the three inorganic realms described above.

Physical geography is a study of the atmosphere, hydrosphere, and lithosphere in relation to the biosphere.

NATURAL SYSTEMS IN PHYSICAL GEOGRAPHY

In what sense is physical geography concerned with natural systems? Explain.

What is the energy source for most natural systems?

NATURAL SYSTEMS IN PHYSICAL GEOGRAPHY 1

Activities and changes that go on within the life layer can best be analyzed in terms of flow systems of matter and energy. A flow system consists of connected pathways through which matter, or energy, or both move continuously.

Most natural systems that comprise physical geography are powered by solar energy.

AN ENVIRONMENTAL SCIENCE

How does physical geography illuminate problems of human survival on the planet?

How is physical geography related to environmental science?

AN ENVIRONMENTAL SCIENCE 1

Physical geography makes assessments of the ability of each kind of natural environment to support human life by furnishing fresh water and food.

Environmental science, the study of interaction between humans and their environment, is closely interwoven with physical geography.

A PLAN FOR STUDY

What sequence of major topics is followed in our study of physical geography?

A PLAN FOR STUDY 2

Our study of physical geography begins with the atmosphere and its climate, followed by study of the hydrosphere and lithosphere, processes that shape the earth's landforms, soil science, ecology, and biogeography.

CHAPTER 1

OUR ROTATING PLANET

PAGE

PROOF OF EARTH'S SPHERICITY

What visual evidence is available to humans that the earth's surface is spherical?

PROOF OF EARTH'S SPHERICITY 3

Evidences of earth curvature: disappearing ships, earth's circular shadow on moon during eclipse, change in altitude of North Star (Polaris).

MEASURING THE EARTH'S CIRCUMFERENCE

Describe Eratosthenes' method of measuring the earth's circumference. What method did the Arabs use to achieve the same information?

MEASURING THE EARTH'S CIRCUMFERENCE 4

Eratosthenes' method used angle of sun's rays from vertical multiplied by ground distance. Arabs used change in angle of star above horizon.

GRAVITY IN THE ENVIRONMENT

Define *gravity*. How is gravity an environmental factor? How does it govern the arrangement of layers of air, water, and rock?

Define *gravitation*.

In what way can gravity be used to prove the earth's sphericity?

GRAVITY IN THE ENVIRONMENT 5

Gravity is an environmental factor: causes layering of air, water, and rock by density. Gravity powers streams and glaciers.

Gravitation: mutual attraction of any two masses.

Constancy of gravity over the earth's surface is a good proof of sphericity.

EARTH AS AN OBLATE ELLIPSOID

Describe and explain Richer's discovery in 1671 of rate change of clock taken to the equator.

Define *oblate ellipsoid*.

EARTH AS AN OBLATE ELLIPSOID 5

Richer's discovery: clock ran slower when taken to equator. Gravity is weaker there, because surface is farther from earth's center of gravity. Oblateness of earth suggested.

Oblate ellipsoid: geometric solid resembling a flattened sphere, with polar axis shorter than equatorial diameter.

What form has the earth's cross section through the poles? through the equator?

Cross section of earth through poles is an ellipse; through equator is a circle.

Describe the form of an *ellipse*.

Ellipse: circle deformed by flattening to produce an oval figure with one maximum diameter and one minimum diameter.

What is the cause of the earth's oblateness?

Earth's oblateness is caused by rotation on its axis.

Define *oblateness*.

Oblateness: ratio of difference between length of polar axis and length of equatorial diameter of an oblate ellipsoid to the equatorial diameter, expressed as a simple fraction.

What is the approximate oblateness of the earth?

Earth's oblateness: approximately 1/300.

Give in rough figures the earth's average diameter and circumference.

Rough figures of earth dimensions: average diameter: 13,000 km (8000 mi); average circumference: 40,000 km (25,000 mi.)

GREAT CIRCLES AND SMALL CIRCLES

GREAT CIRCLES AND SMALL CIRCLES 6

Define a *great circle.*

Great circle: circle formed by passing a plane through the exact center of a perfect sphere.

Explain how great circles are used in navigation.

Great circle courses are used in navigation because they show the shortest distance between any two points on the earth's surface.

Define a *small circle.*

Small circle: circles produced by planes passing through a sphere anywhere except through the center.

MERIDIANS AND PARALLELS

MERIDIANS AND PARALLELS 6

Define *geographic grid.*

Geographic grid: complete network of parallels and meridians on the surface of the globe, used to fix locations of points.

Define *meridians* of longitude.

Meridians of longitude: halves of great circles, ending at either pole.

Give the important properties of meridians.

Properties of meridians: (1) true north-south lines; (2) spaced farthest apart at equator; (3) converging toward either pole; (4) infinite in number.

Define *parallels* of latitude.

Parallels of latitude: small circles produced by passing planes through earth parallel with plane of equator.

Give the important properties of parallels.

Properties of parallels: (1) always parallel with one another; (2) are true east-west lines; (3) intersect meridians at right angles; (4) are all small circles, except equator; (5) infinite number can be drawn.

LONGITUDE

Define *longitude.* What is the maximum value of longitude?

Define *prime meridian.* What meridian is accepted as the prime meridian?

How is longitude written, including minutes and seconds of arc?

How many kilometers (miles) are equivalent to one degree of longitude at the equator?

LONGITUDE 7

Longitude: arc of a parallel east or west of prime meridian. Measures distance in degrees east or west. Maximum is 180°.

Prime meridian: designated reference meridian of zero longitude. *Greenwich Meridian* is universally accepted.

Example of longitude: long. 77° 03' 41"W.

Length of a degree of longitude at the equator: 111 km (69 mi), approximately.

LATITUDE

Define *latitude.* What is the maximum value of latitude?

How is latitude written, including minutes and seconds?

What is the approximate equivalent in km (mi) of one degree of latitude?

How does earth's oblateness affect length of a degree of latitude?

LATITUDE 7

Latitude: arc of a meridian between a given place and the equator, measured north or south. Maximum value is 90°.

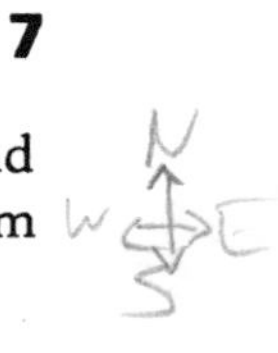

Example of latitude: lat. 34° 10' 31"N.

Length of a degree of latitude averages about 111 km (69 mi).

Oblateness of earth causes a degree of latitude to be longer near poles than near equator. Difference is about 1.1 km (0.7 mi).

NAUTICAL MILE

Define *knot.* In what context is the knot used today?

Define *nautical mile.* Give approximate equivalents in km and statue miles.

NAUTICAL MILE 7

Knot: speed of 1 nautical mi per hour. Used in air navigation and meteorology.

Nautical mile: length of one minute of arc of earth's equator (approx.); 1 nautical mi = 1.85 km = 1.15 mi.

MAP PROJECTIONS

Define *map projections.*

MAP PROJECTIONS 8

A *map projection* is an orderly system of parallels and meridians used as a basis for drawing a map on a flat surface.

ROTATION OF THE EARTH

What is the period of earth rotation? What units of time are used?

Define *earth rotation.*

ROTATION OF THE EARTH 8

Period of *earth rotation* on its axis: one *mean solar day* of 24 mean solar hours.

Earth rotation: spinning of earth on its axis.

Define *mean solar day.*

Mean solar day: time interval required for earth to make one turn with respect to the sun, i.e., from one solar noon to the next.

In what direction does the earth rotate? Explain in three ways.

Direction of rotation: (1) counterclockwise; (2) eastward; (3) opposite to apparent motion of sun, moon, planets, and stars.

What is the approximate speed of travel of a point on the earth's equator? on the 60th parallel?

Speed of surface travel due to rotation: at equator, 1700 km/hr (1050 mph); at 60° lat., 850 km/hr (525 mph).

PROOF OF THE EARTH'S ROTATION

PROOF OF THE EARTH'S ROTATION 8

What two major theories were in conflict concerning the earth's motion with respect to other celestial objects?

Ptolemaic system held that earth is center of universe; Copernican system, that earth and other planets revolve about sun.

What simple apparatus can be used to prove earth rotation on an axis? How does its motion prove rotation?

Foucault pendulum proves earth rotation because hourly rate of change of its direction of swing is always proportional to sine of latitude.

ENVIRONMENTAL EFFECTS OF EARTH ROTATION

ENVIRONMENTAL EFFECTS OF EARTH ROTATION 9

State some important environmental effects of earth rotation.

Environmental effects of earth rotation include daily (diurnal) life rhythms of light, heat, humidity, air motion. Rotation sets up deflective effect (Coriolis effect) on winds and ocean currents; produces lunar tides and tidal currents.

EARTH REVOLUTION

EARTH REVOLUTION 10

Define *orbit, revolution,* and *year.*

Motion of earth in its *orbit,* or travel path, around the sun is *revolution.* The *year* is time interval required for one period of revolution.

How is the *tropical year* defined? Give the length of this year.

Tropical year: one circuit in orbit with respect to sun; i.e., from one vernal equinox to the next. Year length is $365\frac{1}{4}$ days.

What is the direction of earth revolution in its orbit?

Direction of revolution is counterclockwise as viewed from point in space above N. pole.

PERIHELION AND APHELION

PERIHELION AND APHELION 10

What is the average (mean) distance separating earth and sun?

Mean distance between earth and sun: 150 million km (93 million miles).

What is *perihelion?* When does it occur? What is the distance to sun at perihelion?

Perihelion: the point on the earth's orbit at which the earth is closest to the sun, occurs on January 3; distance is $147\frac{1}{2}$ million km ($91\frac{1}{2}$ million miles).

What is *aphelion?* When does it occur?

Aphelion: point on earth's orbit when earth is farthest from sun, occurs on July 4.

Do perihelion and aphelion cause the climatic seasons on earth?

Varying earth-sun distance in orbit is not cause of climatic seasons. Effect too slight to detect.

TILT OF EARTH'S AXIS

TILT OF EARTH'S AXIS 10

Define *plane of the ecliptic.*

Plane of the ecliptic: plane in which earth's orbit lies; in which earth revolves.

What is the angular relationship of earth's axis to the plane of the ecliptic?

Earth's axis always tilted at angle of $23\frac{1}{2}°$ from perpendicular to plane of ecliptic. (Angle is $66\frac{1}{2}°$ with respect to plane of ecliptic.)

What is the angular relation of the earth's axis to fixed stars (space coordinates)?

Earth's axis always points to Polaris (North Star); for this reason the axis undergoes a constantly changing angle with respect to sun.

SOLSTICE AND EQUINOX

SOLSTICE AND EQUINOX 11

Define *summer solstice.* Give dates.

Summer solstice: date on which axis tilt is maximum toward sun. June 21, 22.

Define *winter solstice.* Give dates.

Winter solstice: date on which axis tilt is maximum away from sun. Dec 21, 22.

Define *equinoxes.* When does vernal equinox occur? When does autumnal equinox occur?

Equinoxes: vernal, Mar. 20, 21; *autumnal,* Sept. 22, 23. Dates on which axis tilts neither toward nor away from sun.

THE EQUINOXES

THE EQUINOXES 12

Define *circle of illumination.* What lies on each side of the circle of illumination?

Circle of illumination: a full circle dividing globe into daylight and darkness hemispheres.

Define *subsolar point.* What is the angular relation of sun's rays to earth's surface at this point?

Subsolar point: point on globe where sun's rays are perpendicular to earth's surface (sun in zenith at noon).

Describe conditions at *equinox.* What is the position of the circle of illumination? Where is the subsolar point located?

Equinox conditions: circle of illumination passes through both poles; subsolar point on equator.

What is meant by *solar noon?*

Solar noon at a given place is the instant when the sun reaches highest point in sky and the shadow of a vertical rod points exactly toward geographic north.

What is the *sun's noon altitude?*

Sun's noon altitude: vertical angle of sun above the nearest point of the horizon (south point in northern hemisphere).

How is the sun's noon altitude related to latitude at equinox?

Altitude of sun at noon is equal to the *colatitude* at equinox (but only at equinox).

Define *colatitude.*

Colatitude: angle (arc) equal to 90° minus the latitude of the observer's place.

Describe the apparent path of the sun in the sky at equinox for various latitudes in both hemispheres.

Sun's path in sky at equinox: sun rises exactly at 6 a.m. (local time) due east, sets at 6 p.m. due west, reaches highest point in sky at noon; path of sun lies in plane slanting at angle with respect to observer's horizon equal to colatitude; this angle becomes lower as observer travels poleward, higher with travel toward equator; sun's path is perpendicular to horizon plane at equator, then slants down into northern sky in southern hemisphere.

THE SOLSTICES 15

Describe conditions at either solstice in terms of the position of the circle of illumination.

Solstice conditions: circle of illumination touches arctic circle ($66\frac{1}{2}$°N lat.) and antarctic circle ($66\frac{1}{2}$°S lat.)

Define *arctic circle.* Give latitude.

Arctic circle: parallel of latitude at $66\frac{1}{2}$°N.

Define *antarctic circle.* Give latitude.

Antarctic circle: parallel of latitude at $66\frac{1}{2}$°S.

At winter solstice what area is in continual darkness? What area has continual daylight? Be specific as to latitude. Where is the subsolar point located?

Winter solstice: area poleward of arctic circle is in continual darkness; area poleward of antarctic circle is in continual daylight. Subsolar point lies on tropic of capricorn, $23\frac{1}{2}$°S.

Define *tropic of capricorn.* Give latitude.

Tropic of capricorn: parallel of latitude at $23\frac{1}{2}$°S.

At summer solstice, what area is continually in darkness? in daylight? Be specific. Where is the subsolar point located?

Summer solstice: exchange "south" for "north", "arctic" for "antarctic" in winter solstice description. Subsolar point on tropic of cancer, $23\frac{1}{2}$°N.

Define *tropic of cancer.* Give latitude.

Tropic of cancer: parallel of latitude at $23\frac{1}{2}$°N.

Define *astronomical seasons.* What events define the seasons?

Astronomical seasons: spring, summer, autumn, winter; determined by occurrences of equinoxes and solstices.

Describe conditions at winter solstice. Give all pertinent facts.

Conditions at winter solstice: (1) circle of illumination is tangent to both arctic and antarctic circles; (2) night is longer than day in N. hemisphere, reverse in S. hemisphere; (3) inequality in lengths of day and night increases with latitude; (4) between arctic circle and N. pole night lasts 24 hours; between antarctic circle and S. pole day lasts 24 hours.

Describe conditions at summer solstice.

Conditions at summer solstice: Repeat statements (above) for winter solstice, but substitute "S." for "N."and "N." for "S.". Substitute "antarctic" for "arctic" and vice versa.

SEASONAL CYCLE OF SUN'S DECLINATION

Define *sun's declination.*

Describe the seasonal cycle of the sun's declination. When is the rate of change most rapid?

SEASONAL CYCLE OF SUN'S DECLINATION 16

Sun's declination: latitude of that parallel on which the subsolar point is located at any given instant.

Sun's declination ranges from $23\frac{1}{2}°$S to $23\frac{1}{2}°$N (solstices) and is 0° at equinox. Rate of change is most rapid near either equinox; very slow near either solstice.

TIME

Why are global time relationships important in our modern world?

TIME 16

Because of instantaneous communication by electronic devices and the great rapidity of travel by aircraft, a standard system of relating time to longitude is absolutely essential.

LONGITUDE AND TIME

Define *noon meridian.* Is this meridian real or imaginary?

Define *midnight meridian.* What position does it occupy with respect to the noon meridian?

What is the rate of travel of noon and midnight meridians, given in degrees of longitude per day and per hour; per minute?

Describe a working model of global time, using two discs.

How does the clock time of places east of you compare with your own time?

LONGITUDE AND TIME 17

Noon meridian: imaginary meridian marking occurrence of noon and sweeping westward around globe.

Midnight meridian: imaginary meridian marking midnight; lies opposite to noon meridian.

Rate of travel of noon and midnight meridians: 360° longitude per 24 hr; 15° long. per hr; 1° per 4 min. These figures relate longitude to time.

Working model of global time relationships: disc marked in hour circles turns with respect to disc marked in longitude.

Rule: places east of you have a later hour; places west of you have an earlier hour.

LOCAL TIME

Define *local time.*

LOCAL TIME 17

Local time: mean solar time based on the local meridian passing through a given point on the globe.

STANDARD TIME

Why is it necessary to use standard time, rather than local time?

Define *standard time.*

Define *standard meridians.* How are they spaced apart?

STANDARD TIME 18

Standard time system introduced when telegraph brought about instantaneous communication; was needed for railroad operations.

Standard time: global time system based on standard meridians.

Standard meridians: meridians separated by 15° and occurring in multiples of 15.

Describe standard time zones. How wide is a time zone?

Time zones are north-south standard time belts, each referred to a standard meridian; each spans 15° of longitude; time in each differs by whole hour units from Greenwich time.

STANDARD TIME IN THE UNITED STATES

Name the United States time zones in order from east to west. Give the standard meridian for each.

How are time zone boundaries adjusted for convenience? Give examples of boundaries.

STANDARD TIME IN THE UNITED STATES 18

United States time zones: Eastern, 75°; Central, 90°; Mountain, 105°, Pacific, 120°; Alaskan and Hawaiian, 150°; Bering, 165° (all W. longitude).

Time zone boundaries vary widely from meridian dividing time zones equally. State boundaries often used as time boundaries. Mountain divides, lakes, also used.

DAYLIGHT SAVING TIME

Define *daylight saving time.*

Name some countries using daylight saving time throughout the entire year.

Compare the U.S. national period of daylight saving time with cycle of sun's declination. (See Figure 1.22, pg. 16.)

DAYLIGHT SAVING TIME 18

Daylight saving time: clocks run one hour ahead of standard time.

Some countries use daylight time through entire year. Examples: Spain, France, Netherlands, Belgium, Soviet Union.

Daylight time period in U.S. does not fit cycle of sun's declination; begins long after vernal equinox but ends long after autumnal equinox.

WORLD TIME ZONES

Describe the system of world time zones.

Do all nations conform to the world time zone systems? Explain.

WORLD TIME ZONES 19

World time zone system has 24 zones; 12 with fast time (east longitude), 12 with slow time (west longitude).

Some nations use $7\frac{1}{2}°$ meridians for their standard time, differing by one-half hour from neighboring zones and Greenwich time.

INTERNATIONAL DATE LINE

What standard time applies to the meridian of 180°?

What rule applies to calendar days on either side of the meridian of 180°?

INTERNATIONAL DATE LINE 19

Standard time of meridian of 180° is both 12 hrs. fast and 12 hrs. slow; time difference at 180° is 24 hours, a full calendar day.

Calendar day on the west (Asiatic) side is one day ahead (fast) of that on east (American) side.

What is the *International Date Line?*

International Date Line: the 180° meridian on longitude, together with deviations east and west of that line, forming the boundary between adjacent time zones that are 12 hrs. fast and 12 hrs. slow with respect to Greenwich standard time.

CHAPTER 1—SAMPLE OBJECTIVE TEST QUESTIONS

A. MATCHING

1. latitude
2. equator
3. ellipsoid
4. prime meridian
5. equinox

a. ______ arc of parallel
b. ______ Greenwich
c. ______ arc of meridian
d. ______ small circle
e. ______ great circle
f. ______ vernal
g. ______ oblate

B. MULTIPLE CHOICE

1. A parallel of latitude

______ a. is half of a circle.

______ b. runs north-south.

______ c. is a full circle.

______ d. is in a plane parallel with the earth's axis.

2. Richer observed that at the equator his pendulum clock beat

______ a. slower than

______ b. faster than

______ c. at exactly the same rate as

______ d. first slower, then faster

than the same clock when it was situated in France.

3. Standard time meridians are

______ a. separated by 15 degrees of longitude.

______ b. separated by 15 degrees of latitude.

______ c. in constant motion around the globe.

______ d. allowed to deviate to follow state boundaries.

4. The Foucault pendulum

______ a. swings entirely free of its supporting structure.

______ b. changes direction of swing more rapidly at higher latitudes.

______ c. changes direction 360° in 24 hours at north pole.

______ d. (all three of the above are correct.)

C. COMPLETION

1. The mutual attraction of any two masses is____.
2. A half-circle connecting the two poles is called a ____.
3. A circle produced when a plane is passed through the center of a sphere is a ____.
4. The ______________________________ is a unit of distance measurement approximately equal to one minute of arc of the earth's equator.
5. The period of rotation of the earth with respect to the sun is the ______________________________ day.
6. Motion of the earth in its travel path, or orbit, about the sun is termed______________________________.

CHAPTER 2

THE EARTH'S ATMOSPHERE AND OCEANS

PAGE

Name and define the four material realms that comprise the total global environment.

Four great material realms comprise the total global environment: *atmosphere, hydrosphere, lithosphere, biosphere*. Atmosphere is gaseous realm; hydrosphere is water realm; lithosphere is mineral (rock) realm; biosphere is organic realm.

Use the interface concept to relate the atmosphere to the land surface and ocean surface.

Land surface is an interface with atmosphere across which matter (water) and energy are exchanged. This interface is the life layer of humans. Ocean surface is a similar interface, important because of its control over global climate.

What is *meteorology?*

Meteorology: science of the atmosphere.

What is *physical oceanography?*

Physical oceanography: physical science of the oceans.

COMPOSITION OF THE ATMOSPHERE

COMPOSITION OF THE ATMOSPHERE 23

Of what four major gases is the atmosphere composed? Give percentages of each.

Pure dry air, a mixture of gases:

What part does nitrogen play in chemical processes? Is it used by plants?

Nitrogen, 78%. Largely inactive chemically. Some used by plants.

How active chemically is oxygen?

Oxygen, 21%. Highly active chemically, through oxidation. Essential to life processes.

What is the importance of argon to life?

Argon, almost 1%. Inert.

What role is played by carbon dioxide? Is it important to life? Explain.

Carbon dioxide. 33/1000% (0.033%). Absorbs radiant heat. Used by plants in photosynthesis to make carbohydrate.

ATMOSPHERIC PRESSURE

Define *atmospheric pressure.*

State normal atmospheric pressure in kg/sq cm (lb/sq in.).

Describe the Torricelli experiment? What does it prove? Give the height of mercury column in centimeters and inches.

What is a *barometer?* What does it measure?

What is a *mercurial barometer?*

What is standard sea-level pressure in centimeters (inches) of mercury?

Define *millibar.* How is it abbreviated? What is standard sea-level pressure in millibars?

How many millibars are equal to one cm of mercury?

Describe the workings of the *aneroid barometer.*

ATMOSPHERIC PRESSURE 23

Atmospheric pressure: pressure exerted by the atmosphere because of the force of gravity acting upon the overlying column of air.

Normal atmospheric pressure: 1 kg/sq cm (15 lb/sq in.).

Torricelli experiment. Mercury column in tube stands at height of 76.0 cm (30 in.) at sea level.

Barometer: instrument that measures atmospheric (barometric) pressure.

Mercurial barometer: barometer using Torricelli principle.

Standard sea-level pressure: 76.0 cm (29.2 ins.).

Millibar (mb): pressure unit used in modern meteorology. Standard sea-level pressure = 1013.2 mb.

Equivalents: 1 cm mercury column = 13.3 mb (1 in. = 34 mb).

Aneroid barometer: barometer using hollow metal box, partly evacuated of air, with flexible diaphragm that moves in response to changing air pressure.

VERTICAL DISTRIBUTION OF ATMOSPHERIC PRESSURE

Where is the air most dense? How does density change upward?

How does pressure change with increase in altitude? Give a rate of decrease in terms of fractional drop per unit of altitude.

Describe some physiological and environmental effects of reduced atmospheric pressure.

VERTICAL DISTRIBUTION OF ATMOSPHERIC PRESSURE 24

Air density greatest at surface, decreases rapidly upward.

Pressure decrease with altitude: pressure falls 1/30 for each 275 m (900 ft) altitude increase.

At high altitudes humans experience altitude sickness, shortness of breath. Supplementary oxygen needed above 5400 m (18,000 ft). Boiling point drops with altitude increase; pressure cookers advantageous at high levels.

TEMPERATURE STRUCTURE OF THE LOWER ATMOSPHERE

What is the *troposphere?* How does temperature change in this layer.

What is the *environmental temperature lapse rate?* Give rate in C° per 1000 m.

TEMPERATURE STRUCTURE OF THE LOWER ATMOSPHERE 25

Troposphere: lowermost layer in which temperature falls steadily with increasing altitude (on the average).

Environmental temperature lapse rate: 6.4 C°/1000 m ($3\frac{1}{2}$ F°/1000 ft), average value.

Define *tropopause*. Between what two atmospheric layers is the tropopause a boundary? At what altitude is it located?

Tropopause: level at which temperature no longer falls with increased altitude; it is the boundary between troposphere and stratosphere, located at about 14 km (9 mi) altitude in midlatitudes.

Is the tropopause higher over the equatorial zone than over the polar zones? Explain.

Tropopause in winter is higher (17 km) and colder (–70°C) over equatorial zone than over polar zones (9 km, –58°C).

What is the *stratosphere?* Where is it located? How does temperature change upward within the stratosphere?

Stratosphere: layer above the tropopause. Temperature increases slowly upward in stratosphere.

What causes the warming of air temperature within the stratosphere?

Absorption of solar energy by ozone results in warming of the stratosphere.

What is the *stratopause?* What layer lies above it?

Stratopause: upper limit of the stratosphere, located at 50 km.

Mesosphere: cold atmospheric layer above the stratosphere between 50 and 80 km

What is the *mesopause?*

Mesopause: upper limit of the mesosphere, at about 80 km altitude.

What layer lies above the mesopause?

Thermosphere: atmospheric layer of rising air temperature above the mesopause.

THE TROPOSPHERE AS AN ENVIRONMENTAL LAYER

THE TROPOSPHERE AS AN ENVIRONMENTAL LAYER 26

What ingredients, besides the gases of pure dry air, does the troposphere contain?

Troposphere contains most of the water vapor, clouds, dust, and weather disturbances.

What is *water vapor?* Is it a gas or a liquid? What happens when water vapor condenses? Can water vapor absorb radiant heat?

Water vapor: the gaseous form of water. Water vapor condenses to form clouds, to yield precipitation. Water vapor absorbs radiant heat energy.

What role does atmospheric dust play in weather processes? Where does this dust come from? Name four sources.

Dust particles serve as nuclei of condensation. Atmospheric dust: derived from land surface, ocean surface, volcanoes, meteors, industrial processes.

THE OZONE LAYER-A SHIELD TO EARTHLY LIFE 27

What is the *ozone layer?* Where is the ozone layer located? How is ozone generated in this layer?

Ozone layer: atmospheric layer within stratosphere; is rich in ozone gas produced by action of ultraviolet rays on oxygen molecules.

Define *ozone.* Give the chemical composition. How is it different from ordinary atmospheric oxygen?

Ozone: oxygen gas molecule consisting of three atoms of oxygen (O_3), instead of usual two (O_2).

Describe the shielding action of ozone. What effect have ultraviolet rays upon organic matter?

Shielding action of ozone layer: absorbs ultraviolet (UV) rays destructive to organic tissues, genetic materials.

In what way is the ozone layer threatened by the release of synthetic compounds into the lower atmosphere?

Release of Freons, also called *halocarbons,* at the earth's surface threatens the integrity of the ozone layer because these compounds absorb ultraviolet radiation; they are decomposed to release chlorine, which destroys ozone molecules.

What are *halocarbons?*

Halocarbons: synthetic compounds containing carbon, fluorine, and chlorine atoms. (Fluorine and chlorine belong to a class of elements known as halogens.)

What are the results of studies of the effects of halogens on the ozone layer? What action has been taken?

Scientific investigations indicate that substantial depletion of the ozone layer can be expected, even through use of halocarbons in spray cans has been banned since 1976.

THE EARTH'S MAGNETIC ATMOSPHERE

THE EARTH'S MAGNETIC ATMOSPHERE 28

Describe the earth's *external magnetic field.* What name is given to it?

External magnetic field, or, *magnetosphere:* external portion of the earth's magnetic field, consisting of lines of magnetic force.

What name is given to the outer boundary of the magnetosphere? What force shapes that outline?

Magnetopause: outer boundary of the magnetosphere, shaped by pressure of the solar wind.

What is the *solar wind?*

Solar wind: flow of electrons and protons (charged particles) emanating from the sun.

What environmental role does the magnetosphere play?

Magnetosphere shields earth from ionizing radiation capable of destroying life.

What is *ionizing radiation?*

Ionizing radiation: electromagnetic radiation of very short wavelengths, such as X rays and radioactivity.

What is the *Van Allen radiation belt?*

Van Allen radiation belt: ring of intense ionizing radiation surrounding earth.

SOLAR FLARES AND MAGNETIC STORMS

SOLAR FLARES AND MAGNETIC STORMS 29

What causes a *magnetic storm?*

Magnetic storm: caused by the arrival of energetic particles from an intense solar flare.

Define solar flare.

Solar flare: emissions of ionized hydrogen gas from the vicinity of a sunspot.

Define the *ionosphere.*

Ionosphere: a layer of ionized atmospheric gases that sets in at 50 km (30 mi) and extends to 1000 km (600 mi).

Describe the *aurora borealis* or *aurora australis.*

The *aurora borealis, aurora australis:* a product of a magnetic storm, seen at high northern or southern latitudes. The aurora takes the form of light bands, rays or draperies continually shifting in pattern and intensity in the night sky.

OCEANS AND THE HUMAN ENVIRONMENT

Describe the climatic role played by the oceans. How do they regulate heat? What gas do they supply to the atmosphere in large amounts. How are the oceans a resource for humans? How are oceans a hazard to humans? How do ocean waves affect the lands?

OCEANS AND THE HUMAN ENVIRONMENT 30

Oceans in climatic role. Oceans hold enormous quantity of heat in storage, moderating climate. Oceans supply water vapor to atmosphere. Oceans sustain life valuable as food to humans. Ocean offers environmental hazards. Ocean waves shape borders of lands.

THE WORLD OCEAN

Define *world ocean.*

What is the extent of the world ocean in percent of area?

What is the average depth of the world ocean?

What is the volume of the world ocean in billion cu km? in terms of percent of world's free water?

Define *hydrosphere.* What states of water does it include?

Compare world ocean with atmosphere in terms of each of the following points:

Compressibility and density. Where is the atmosphere densest? How does ocean density change with depth?

Upper boundary: Distinct or indistinct?

Most active layer: Lowest or uppermost?

Speed of motion: Rapidly or slowly?

Describe activities at ocean-atmosphere interface:

How do atmospheric motions produce motion in oceans?

THE WORLD OCEAN 30

World ocean: combined ocean bodies and seas of globe.

Extent: 71% of global area.

Average depth: 3800 m (12,500 ft) for major oceans.

Volume comparison of world ocean and atmosphere: 1.4 billion cu km (317 million cu mi), or 97% of world's free water.

Hydrosphere: general term encompassing all free water of earth in gaseous, liquid, and solid states.

Comparison of world oceans and atmosphere

Atmosphere:	World ocean:
Gas, easily compressed, is much denser at bottom.	Not easily compressed. Density changes only slightly with depth.
No distinct upper boundary.	Sharply defined upper surface.
Lowest layer (troposphere) is most active.	Upper layer is most active.
Atmosphere moves easily, rapidly.	Ocean water moves slowly.

Ocean-atmosphere interface

Atmospheric motions (winds) drive ocean motions (currents).

What gases enter the ocean from the atmosphere?

Oxygen and carbon dioxide enter ocean from atmosphere.

Is heat exchanged? In what direction?

Heat enters ocean from atmosphere, and vice versa.

Is water vapor transferred?

Water vapor enters atmosphere from oceans.

COMPOSITION OF SEAWATER

What are the most important salts in seawater, in order of dissolved weight?

COMPOSITION OF SEAWATER 31

The most important salts in seawater are sodium chloride ($NaCl$), magnesium chloride ($MgCl_2$), sodium sulfate (Na_2SO_4), calcium chloride ($CaCl_2$), and potassium chloride (KCl).

LAYERED STRUCTURE OF THE OCEANS

Describe the temperature layering of the oceans. What layers are present in low latitudes and midlatitudes? Are layers present in arctic and antarctic latitudes?

Define *thermocline*. Where is the thermocline?

Describe the oxygen layering of the ocean. Why is there a layer of minimum oxygen?

LAYERED STRUCTURE OF THE OCEANS 31

Temperature layering. Warm surface layer (low latitudes and midlatitudes), thermocline, cold deep layer. Arctic and antarctic latitudes: cold from surface down.

Thermocline: water layer in which temperature changes rapidly in the vertical direction.

Oxygen layering. Oxygen-rich surface layer above, oxygen minimum layer below. Oxygen depleted by organic activity.

FLOW SYSTEMS OF MATTER AND ENERGY 33

MATTER AND ENERGY

Of what two basic components are all of the physical earth realms composed?

Define *matter.*

Define *energy.*

MATTER AND ENERGY 33

Matter and *energy* constitute the basic components of the four physical earth realms: atmosphere, hydrosphere, lithosphere, and biosphere.

Matter: tangible substance occupying space and responding to *gravitation* (the mutual attraction between any two aggregates or particles of matter).

Energy: the ability to do work, for example, through the motion of matter or through changes in state of matter.

STATES OF MATTER

In what three *states* does matter exist?

Define *gas.* Can a gas be easily compressed?

STATES OF MATTER 33

Matter exists in three states: *gaseous state, liquid state, solid state.*

Gas: substance that expands to fill uniformly any small container, has low density, and is readily compressible. Gas molecules travel at high speeds.

Define *liquid.* Does a liquid move freely? Has it a free upper surface?

Liquid: substance that flows freely in response to unbalanced forces and maintains a free upper surface.

What is a *fluid?*

Both gases and liquids are classed as *fluids* because they respond freely to gravity by flowage and form a natural layering in order of increasing density from top to bottom.

What is a *solid?* Is a solid easily compressed? Does it possess strength?

Solid: substance that resists change of shape and volume, has strength, and yields by sudden breakage.

FORMS OF ENERGY

FORMS OF ENERGY 34

List the commonly recognized forms of energy.

Commonly recognized forms of energy are mechanical, heat, chemical, electrical, and nuclear.

What is *mechanical energy?* In what two forms does it occur?

Mechanical energy: energy associated with the motion of matter as aggregates of molecules. Two forms are *kinetic energy* and *potential energy.*

What is *kinetic energy?* How does quantity of kinetic energy related to speed?

Kinetic energy: the ability of a mass in motion to do work. Kinetic energy varies in proportion to the square of the speed of the moving mass and directly as the quantity of mass.

What is *potential energy?* How is it related to gravity and elevation?

Potential energy: energy of position, capable of transformation into kinetic energy under the force of gravity. For a given unit of mass, potential energy is directly proportional to height above a reference surface and to the acceleration of gravity.

Describe *wave motion* as a form of energy.

Mechanical energy can be transmitted through a substance as *wave motion* in which kinetic energy is passed along as an impulse. Examples: sound waves, earthquake waves.

What is *sensible heat*? What form of energy does it represent?

Sensible heat, measured be the thermometer, is an internal form of kinetic energy sustained by molecular motion within a substance.

How does sensible heat travel through or within a substance?

Sensible heat may travel by *conduction* from one molecule to another or by *convection*, a process of mixing by currents within a fluid.

What is *latent heat?*

Latent heat: form of energy stored within a substance and capable of transformation into sensible heat when a change of state occurs.

Describe *electromagnetic energy.* How is it related to *radiation?* What is the *electromagnetic spectrum?*

Electromagnetic energy: form of energy transport capable of traveling through space and transparent substances in a straight path at the uniform speed of light as a form of *radiation.* It can be visualized as a spectrum of waves of a wide range of lengths—the *electromagnetic spectrum*—including ultraviolet rays, X rays, gamma rays, light rays, infrared rays, and longer waves.

What is *chemical energy?*

Chemical energy: energy absorbed or released by matter when chemical reactions take place. Example: energy absorbed by organic molecules synthesized in plants.

What is *electrical energy?*

Electrical energy: flow of free electrons through a conductor; as in an electric current, or stored by accumulated electrons as a charge.

What is *nuclear energy?*

Nuclear energy: energy released by the spontaneous disruption of atomic nuclei of radioactive substances.

FLOW SYSTEMS OF MATTER AND ENERGY

FLOW SYSTEMS OF MATTER AND ENERGY 35

Describe a *flow system.*

Flow system: series of interconnected pathways through which energy or matter moves more or less continuously.

What is an *energy flow system?*

Energy flow system: system of flow of energy through a flow path from a point of entry to a point of exit in the *system boundary.*

What is a *material flow system?*

Material flow system: system of flow of matter through pathways in a material structure, including places of temporary storage of matter.

Define *system boundary.*

System boundary: real or arbitrary separating one system from another. Example: the outer wall of a living cell.

Does a material flow system require energy? Explain.

Flow of matter in a material flow system requires the expenditure of energy. Therefore, an energy flow system exists in conjunction with every material flow system.

Can energy be transformed and stored within an energy flow system? Explain.

Within an energy flow system energy can be transformed; for example, radiant energy can be transformed into sensible heat, (*energy transformation*) which can be stored temporarily (*energy storage*).

What is meant by *steady state* of system operation?

Steady state within a system exists when the rate of input equals the rate of output and storage is a constant quantity.

What is an *open system?*

Open system: system requiring both an input and an output through the system boundary.

Are energy systems open systems?

All energy systems are open systems. No system boundary exists that can prevent the movement of energy in or out of the system.

What is a *closed system?*

Closed system: system in which all flow occurs within the system boundary.

Can material flow systems be closed systems? Explain.

Material flow systems can be closed systems, because matter can be cycled within the system boundary. This is possible because an open energy system powers the flow of matter.

CHAPTER 2—SAMPLE OBJECTIVE TEST QUESTIONS

A. MATCHING

1.	gas	a. _______	basal layer
2.	liquid	b. _______	barometer
3.	meteorology	c. _______	clouds
4.	Torricelli	d. _______	temperature change
5.	ozone layer	e. _______	global water
6.	water vapor	f. _______	compressible
7.	hydrosphere	g. _______	free upper surface
8.	troposphere	h. _______	pressure unit
9.	thermocline	i. _______	atmospheric science
10.	millibar	j. _______	stratosphere

B. MULTIPLE CHOICE

1. Within the troposphere

_______a. temperature increases with altitude increase.

_______b. temperature decreases with altitude increase.

_______c. lies the ozone layer.

_______d. no water vapor is present.

2. Atmospheric dust is

_______a. entirely absent in the troposphere.

_______b. concentrated in the ozone layer.

_______c. important as nuclei of condensation.

_______d. concentrated at the tropopause.

3. Within the thermocline

_______a. temperature increases downward.

_______b. temperature holds constant.

_______c. oxygen is concentrated.

_______d. temperature decreases downward.

4. Which of the following is not a fluid?

_______a. liquid water.

_______b. water vapor.

_______c. carbon dioxide gas.

_______d. electric current.

5. An energy flow system

   ________a. may exist as either a closed system or an open system.

   ________b. must have a material flow system in which to operate.

   ________c. can exist only as an open system.

   ________d. cannot store energy within the system.

C. COMPLETION

1. The science of the atmosphere is ____.
2. The most abundant gas of the atmosphere is ______________________; its percentage is ________________%.
3. The second most abundant gas is ____ with ________________%.
4. A gas present in very small amount in the air, capable of absorbing radiant heat is ____________________________.
5. The layer immediately above the tropopause is the ____stratosphere____; below the tropopause is the ____troposphere____.
6. A gas molecule consisting of three oxygen atoms is ____Ozone____; this gas is capable of absorbing ____UV____ radiation.
7. The earth's external magnetic field, also called the ____magnetosphere____, is shaped by pressure of the ____solar winds____.
8. Heat as measured by a thermometer is ____sensible____.

CHAPTER 3

THE EARTH'S RADIATION BALANCE

PAGE

Define *radiation balance.* What forms of energy flow are in balance?

Radiation balance: the condition of balance between solar (shortwave) energy absorbed by earth and longwave energy radiated by earth into space.

THE SUN'S ENERGY OUTPUT

THE SUN'S ENERGY OUTPUT 38

Define electromagnetic radiation. What form of energy in involved? How fast does it travel?

Electromagnetic radiation: wavelike form of energy radiated by any substance possessing heat; travels through space at speed of light, 300,000 km/sec (186,000 mi/sec).

How are waves of the electromagnetic spectrum described?

Entire range, or spectrum, of electromagnetic waves is described in terms of either *wavelength* or *wave frequency.*

Define *wavelength.* What units of wavelength are used?

Wavelength: distance separating one wave crest from the next in any uniform succession of traveling waves. Units of wavelength are *micron* 0.0001 cm (10^{-4} cm), cm. m.

Define *wave frequency.*

Wave frequency: number of waves passing a fixed point per unit of time (per second).

What unit of measure describes wave frequency?

Measure of frequency is the *hertz,* equal to one cycle per sec. One *megahertz* is 1,000,000 cycles per sec.

What form of radiation lies at the short wavelength end of the spectrum?

Gamma rays, wavelengths shorter than 0.03 *angstroms,* occupy short wavelength end of spectrum.

Define *angstrom.*

Angstrom: wavelength unit equal to 0.000,000,01 (10^{-8} cm).

What are x rays?

X rays: wavelengths between 0.03 and 100 angstroms.

Which forms can be described as ionizing radiation? (See Chapter 2, p. 28.)

Both gamma rays and X rays are forms of ionizing radiation.

What are *ultraviolet rays?* What range of wavelengths do they span?

Ultraviolet rays: electromagnetic radiation in the wavelength range 0.2–0.4 microns.

What part of the spectrum consists of *visible light?* Name the colors in order from shortest to longest wavelength.

Visible light: electromagnetic radiation in the wavelength range 0.4–0.7 microns. Colors: violet (shortest), blue, green, yellow, orange, red (longest).

What are *infrared rays?*

Infrared rays: invisible rays can sometimes be felt as "heat rays" from hot object: wavelengths from 0.7–300 microns.

MEASURING INCOMING SOLAR RADIATION

MEASURING INCOMING SOLAR RADIATION 40

What is the *solar constant?* What is its value in calories.

Solar constant: intensity of solar radiation falling on 1 sq cm surface at right angles to sun's rays; value: 1400 watts per square meter.

Define *joule.*

Joule (J): the standard unit of energy (or work).

What is the *watt?*

The *watt (W)* describes the rate at which energy flow in solar rays or through conducting metals.

ENERGY SPECTRA OF SUN AND EARTH

ENERGY SPECTRA OF SUN AND EARTH 40

How does the intensity of energy radiation vary with temperature?

Energy emitted per unit area by a substance increases with increasing temperature.

Define *black body.*

Black body: ideal surface that is a perfect radiator of energy and will absorb all energy falling upon it.

What is meant by *absolute temperature?*

Absolute temperature in *Kelvin degrees* begins with 0° equal to –273°C. Units are of same value as Celsius units.

State the *Stefan-Boltzmann law.*

Stefan-Boltzmann law: total energy radiated per unit surface area of a *black body* varies as fourth power of *absolute temperature,* (°K).

How do the dominant wavelengths of the radiation spectrum relate to surface temperature of a radiating body?

As temperature of radiating surface increases, peak of radiation spectrum moves toward shorter wavelengths.

Compare the radiation spectrum of the sun with that of the earth.

Sun's radiation spectrum resembles that of black body at 6000°K; peak is in visible light region. Earth's spectrum resembles that of black body at 300°K, entirely within infrared region; peak is at about 10 microns.

About what proportion of the total solar energy is in each part of the electromagnetic radiation spectrum?

Sun's spectrum; 9% ultraviolet and shorter wavelengths, 41% visible light, 50% infrared (nearly all in wavelengths shorter than 5 microns).

In what wavelength bands does out-going longwave radiation escape to outer space? What is a *window?*

Much outgoing infrared radiation is absorbed by atmospheric water vapor and CO_2. Certain bands, called *windows,* allow infrared to pass into outer space: 4–6, 8–14, 17–21 microns.

Define *shortwave radiation.* To what radiation source does this term apply?

Shortwave radiation: spectrum of solar electromagnetic radiation, mostly ranging from 0.2 to 3 microns.

Define *longwave radiation.* To what radiation source does this term apply?

Longwave radiation: spectrum of earth radiation in the infrared range, mostly from 3 to 50 microns.

INSOLATION OVER THE GLOBE

INSOLATION OVER THE GLOBE 41

Define *insolation.* What units of measure are used?

Insolation: the interception of solar (shortwave) energy by an exposed surface. Units are langleys per unit of time, e.g., ly/min, ly/yr.

What two factors determine insolation per day at a given place?

Factors determining insolation per day at given point on globe:

How does angle of sun's rays affect insolation? How does this angle change with season?

(1) Angle at which rays strike earth's surface. Angle determined by latitude and season.

How does length of time of exposure affect insolation? How does this time depend upon season?

(2) Length of time of exposure to sun's rays. Depends on season and sun's path in sky.

Describe conditions of insolation at equinox. Where is the subsolar point located at this date? At what latitude is insolation 50% of the equator value? Where is insolation zero? (See Figures 1.15 and 1.17 p. 12–13.)

Equinox conditions: subsolar point over equator, where insolation is the maximum. Decrease with latitude to 50% at 60° latitude; 0% at poles.

How does axial tilt affect insolation? Through what latitude range does the subsolar point shift from one solstice to the next?

Effect of axial tilt on insolation: subsolar point shifts through 47° of latitude from one solstice to the next. This annual cycle of sun's declination has two effects: (1) distributes insolation toward poles; (2) causes seasonal cycle in insolation.

What area of the globe experiences two maxima and two minima in yearly cycle of insolation?

Area between tropics of cancer and capricorn ($23\frac{1}{2}$°N and $23\frac{1}{2}$°S) show two maxima and two minima of insolation.

Describe the change in annual cycle of insolation from middle to high latitudes.

Traced poleward from middle to high latitudes in northern hemisphere, insolation cycle shows greater annual range and higher peak at summer solstice. Highest value occurs at solstice at N pole.

WORLD LATITUDE ZONES

WORLD LATITUDE ZONES 43

Name seven latitude zones in order from equator to poles. Define each zone.

Zones named for convenience in geographical description:

Equatorial zone: 0°–10° N and S; centered on equator.

Tropical zones: 10°–25° N and S; centered on tropics of cancer and capricorn.

Subtropical zones: 25°–35° N and S.

Midlatitude zones: 35°–55° N and S.

Subarctic (Subantarctic) zone: 55°–60° N and S.

Arctic (Antarctic) zone: 60°–75° N and S; centered on arctic and antarctic circles.

Polar zones: 75°–90° N and S.

INSOLATION LOSSES IN THE ATMOSPHERE

What wavelength bands are absorbed in the upper atmosphere? How is this absorption related to the ozone layer?

Define *scattering.* In what way is reflection involved? What particles cause scattering?

Define *diffuse reflection.* What particles cause diffuse reflection?

Define *down-scatter.* In what direction does down-scatter travel?

Define *absorption.* How does energy loss occur? How is energy transformed during absorption? Which gases of the atmosphere are the principal absorbers?

Define reflection. How is reflection different from radiation?

Define *cloud reflection.* Where does the reflected energy go?

Define *albedo.* What units are used? What is the earth's albedo, as an average planetary value?

Summarize insolation losses in atmosphere, giving percentages:

Diffuse reflection to space _______ %

Reflection from clouds to space _______ %

Direct reflection from earth's surface _______ %

Total losses by reflection _______ %

INSOLATION LOSSES IN THE ATMOSPHERE 43

X rays and some ultraviolet energy lost by absorption at altitude of 88 km (55mi). (Review ozone layer, Chapter 3.)

Scattering: turning aside of shortwave solar radiation in all directions by gas molecules of atmosphere; a reflection process.

Diffuse reflection: scattering caused by minute particles of dust and clouds.

Down-scatter: scatter and reflection of all forms directed downward toward the earth's surface.

Absorption: loss of radiant energy to the gas through which the rays are passing. Energy is transformed to heat energy within the absorbing gas. CO_2 and H_2O are principal absorbing gases in lower atmosphere. (Liquid and solid surfaces can also absorb radiant energy.)

Reflection: turning back of rays by a reflecting surface without loss of energy.

Cloud reflection: reflection returned to space by upper surfaces of clouds.

Albedo: percentage of electromagnetic radiation reflected from a surface. Total reflection: 100%. Total absorption: 0%. Earth's albedo: about 32% as planetary average.

Summary of insolation losses in atmosphere:

By diffuse reflection to space	6%
By reflection from clouds to space	21%
By direct reflection from earth's surface	4%
Total losses by reflection	31%

Absorption by gas molecules, dust, clouds ______ %

Absorption by earth's surface ______ %

Total losses by absorption ______ %

Total losses by reflection and absorption ______ %

By absorption by gas molecules, dust, clouds 21%

Absorbed by earth's surface 48%

Total losses by absorption 69%

Total losses by reflection and absorption 100%

LONGWAVE RADIATION

Define *ground radiation.* What wavelength band is involved?

Describe *infrared imagery.*

Define *remote sensing.*

LONGWAVE RADIATION 45

Ground radiation: emission of infrared (longwave) radiation by earth's surface, including land and water surfaces.

Infrared imagery image of longwave radiation emitted from objects at night, pavements and rivers appear bright and moist soil surfaces appear dark due to changes in heat radiation.

Remote sensing: a modern research tenhnology makes use of electormagnetic radiation in various bands over wide range of the total spectrum.

THE GLOBAL RADIATION BALANCE

Define *counterradiation.* What wavelengths are involved? How is the net value of ground radiation calculated?

Describe and explain the *greenhouse effect.*

Summarize flow of energy of all forms leaving ground surface and passing upward into atmosphere:

Longwave radiation, net value ______ %

Mechanical transport upward as sensible heat ______ %

Transport upward as latent heat in water vapor ______ %

Total of above three terms______ %

THE GLOBAL RADIATION BALANCE 46

Counterradiation: longwave radiation of atmosphere returned to earth. Net value of longwave ground radiation is difference between total ground radiation and counterradiation.

Counterradiation of longwave energy, causing warming of lower atmosphere, called *greenhouse effect.*

Summary of energy in all forms leaving ground surface and passing upward into atmosphere:

Longwave radiation (net value) 16%

Mechanical transport upward as sensible heat 10%

Transport upward as latent heat in water vapor (evaporation) 22%

Total 48%

THE GLOBAL RADIATION BALANCE AS AN OPEN ENERGY SYSTEM

How can the global radiation balance be visualized as an open energy system?

What is the energy input of this system?

What happens to shortwave energy absorbed by atmosphere and ground?

How is sensible heat stored in the atmosphere?

How is the concept of *energy subsystems* illustrated by this model?

THE GLOBAL RADIATION BALANCE AS AN OPEN ENERGY SYSTEM 46

Global radiation balance can be visualized as an open energy system in which outer limit of atmosphere forms system boundary.

Shortwave solar energy is the input of the system, but part is reflected back directly into space, exiting the system. Net input is difference between absorbed and reflected shortwave energy.

Shortwave energy absorbed by atmosphere and ground is transformed into sensible heat as storage.

Sensible heat stored in ground gives up energy (1) by transformation to longwave radiation; (2) by upward transport as latent heat in water vapor; (3) by sensible heat conduction and convection.

Two *energy subsystems* are recognized: atmospheric subsystem and ground subsystem. Each subsystem is contained within the larger system.

LATITUDE AND THE RADIATION BALANCE

Albedo

Describe the variation in mean albedo with latitude, as seen in a meridional profile.

Insolation

In what zone is the value of insolation (incoming solar radiation) highest? Explain.

Longwave Radiation

Describe the meridional profile of mean longwave radiation from the earth.

LATITUDE AND THE RADIATION BALANCE 47

Albedo 47

Albedo of earth lowest in low latitudes, increases from midlatitude zone into high latitudes; highest over antarctic and S polar zone.

Insolation 48

Maximum mean annual insolation (solar radiation) is in tropical zones, where deserts prevail and cloudiness is low.

Longwave Radiation 49

Mean longwave radiation from earth is greatest in low latitudes, falls off rapidly poleward of lat. 40° N and S.

NET RADIATION

Define *net radiation.* What forms of energy does it combine? What two energy flows are balanced?

Define *energy surplus.* When does a surplus exist? Is it positive or negative?

Define *energy deficit.* When does a deficit exist? Is it positive or negative?

NET RADIATION 50

Net radiation: difference between all incoming energy and all outgoing energy carried by both shortwave and longwave radiation.

Energy surplus: energy coming in faster than going out; a positive value.

Energy deficit: energy going out faster than coming in; a negative value.

In what latitude belt is the global radiation balance positive? In what latitude belts is it negative? Give latitudes.

Global radiation balance is positive (surplus) between 40 N and 40 S; negative (deficit) poleward of 40 N and S.

Define *meridional transport.* In what direction does meridional transport take place? How does it affect the global energy balance?

Meridional transport: transport of energy or matter across the parallels of latitude, in a north-south direction or the reverse. Carries energy from region of surplus to region of deficit.

SOLAR ENERGY 51

How much solar energy does the earth intercept each year? How does this value compare with present rate of energy consumption?

Planet earth intercepts solar energy at the annual rate of $1\frac{1}{2}$ quadrillion megawatt-hours, a quantity about 28,000 times greater than all energy consumption by humans.

What basic advantages has the use of solar energy over other forms of energy?

Solar energy, captured and used by humans for all purposes, adds no heat load upon the atmosphere and creates no air pollution.

What is meant by direct forms of solar energy?

Direct interception and conversion of solar energy takes two forms: (1) direct absorption of shortwave energy to generate and store sensible heat; (2) conversion of shortwave energy to electrical energy by use of solar cells.

Give examples of indirect or secondary sources of solar energy.

Indirect or secondary sources of solar energy include kinetic energy of wind, waves, and streams. Another secondary source is stored chemical energy in plants, derived originally from solar energy.

What important applications has direct absorption of shortwave energy to produce sensible heat?

Direct absorption of shortwave energy has important immediate uses in space heating and hot-water heating, using small systems limited to individual buildings.

What is the most efficient device to intercept solar energy?

Solar collectors, using systems of tubes through which a fluid is circulated, are the most efficient simple devices to collect shortwave energy.

How can solar energy be used in power plants to generate electricity?

Solar power plants use movable mirrors, called heliostats, to reflect solar rays upon a central point where water is heated to extremely high temperatures.

Explain how electricity can be generated directly from solar rays.

Solar cells of crystalline silicon generate an electric current when exposed to sunlight. Large arrays of solar cells are needed to yield high voltages.

What disadvantage does solar energy have when used to generate electricity?

Because electricity is not generated at night, means of storing electrical power must be provided (e.g., battery systems), or the energy must be converted into hydrogen.

CHAPTER 3—SAMPLE OBJECTIVE TEST QUESTIONS

A. MATCHING

1.	ultraviolet	a. ________	8–13 microns
2.	visible light	b. ________	2 langleys/min
3.	longwave radiation	c. ________	north-south
4.	solar constant	d. ________	reflection
5.	tropical zones	e. ________	3–50 microns
6.	midlatitude zones	f. ________	imagery
7.	albedo	g. ________	10°–25° N,S
8.	window	h. ________	0.4–0.7 microns
9.	remote sensing	i. ________	35°–55° N,S
10.	meridional	j. ________	0.2–0.4 microns

B. MULTIPLE CHOICE

1. Longwave radiation

   ________a. refers to the total solar radiation spectrum.

   ___✓___b. is the spectrum of earth radiation.

   ________c. is mostly in the visible light range.

   ________d. takes place only when the sun is shining.

2. The effect of axial tilt upon insolation is to

   ________a. concentrate insolation in the equatorial zone.

   ________b. keep insolation uniform throughout the year.

   ___✓___c. distribute insolation toward the polar regions.

   ________d. absorb more insolation at the poles than at the equator.

3. Scattering results because of the presence of

   ___✓___a. gas molecules in the atmosphere.

   ________b. cloud particles.

   ________c. dust particles.

   ________d. X rays and ultraviolet rays.

C. COMPLETION

1. The condition of balance between solar energy absorbed by the earth and energy radiated into space by earth is the ___radiation balance___.
2. Interception of solar energy by an exposed surface is ___~~absorption~~ insolation___.

3. The latitude zone between 10°S and 10°N is the equator
zone; between 25° and 35°N is the subtrop
zone; between 55° and 60°N is the subarctic
zone.

4. The percentage of electromagnetic energy reflected from a surface is the albedo
____________________.

5. Longwave radiation from atmosphere returned to the earth is counter radiation
____________________.

CHAPTER 4

HEAT AND COLD AT THE EARTH'S SURFACE

PAGE

Define *thermal environment.* How do heat and cold influence activities and processes of the life layer?

Thermal environment: influence of heat and cold in the life layer upon organisms. Biochemical processes and chemical reactions intensified by higher temperatures, inhibited by cold.

Define sensible heat. In what other form can heat be held?

Sensible heat: heat measurable by a thermometer; an indication of the intensity of kinetic energy of molecular motion within a substance. Differs from latent heat, absorbed or released by changes of state.

MEASUREMENT OF AIR TEMPERATURES

MEASUREMENT OF AIR TEMPERATURES 54

Describe the standard thermometer shelter. How is it constructed?

Standard instrument shelter, 1.2–1.8 m (4–6 ft) off ground. Louvered sides permit ventilation.

Define *maximum-minimum thermometer.* What information does it give? How often is it reset?

Maximum-minimum thermometer: pair of thermometers that show highest and lowest values since last reset. Reset once a day.

Define *mean daily temperature* How is it calculated?

Mean daily temperature: sum of daily maximum and minimum readings divided by two.

Define *mean monthly temperature* How is it calculated? What length of period of record is used?

Mean monthly temperature: mean of daily means for a given month; may be based on record of many years.

Define *mean annual temperature* In what ways can it be calculated?

Mean annual temperature: (1) mean of daily means for a given year, or of many years; (2) mean of monthly means.

Compare *Celsius* and *Fahrenheit* temperature scales. Give conversion data.

Celsius scale (C°): freezing, 0°; boiling, 100°.

Fahrenheit scale (F°): freezing, 32°; boiling, 212°.

($C = \frac{5}{9}(F - 32°)$; $F = \frac{9}{5}C + 32°$)

DAILY CYCLES OF INSOLATION, NET RADIATION, AND AIR TEMPERATURE

DAILY CYCLES IN INSOLATION, NET RADIATION AND AIR TEMPERATURE 55

Describe the daily insolation cycle at equinox.

Daily cycle of insolation: insolation begins at sunrise, 6 a.m. local time; reaches peak at solar noon; declines to zero at sunset, 6 p.m. local time.

How is the insolation cycle different at June solstice and December solstice in middle latitudes?

June solstice: insolation starts earlier (about 4 a.m.); reaches high peak at noon; ends later (about 8 p.m.). December solstice: starts about 8 a.m.; noon value is less than at equinox; ends about 4 p.m.

What instrument is used to measure intensity of solar shortwave radiation (insolation)?

Pyranometer measures insolation, including direct beam and indirect sky shortwave radiation.

Describe the daily cycle of net radiation at equinox.

Daily cycle of net radiation is symmetrical, similar to insolation cycle in form. Surplus starts shortly after sunrise, peaks at noon, becomes a deficit shortly before sunset; deficit persists throughout night.

How is the net radiation cycle different at June solstice and December solstice?

June solstice: positive value (surplus) begins earlier; reaches higher noon value; reaches zero later in day. December solstice: surplus begins later; noon peak is lower; ends earlier.

How does the total daily net radiation for a day in June compare with that for a day in December?

In June, net radiation shows large daily surplus, small deficit; in December, deficit may exceed the surplus to give net total negative value.

Describe the daily cycle of air temperature at a midlatitude location.

Daily cycle of air temperature in midlatitude location: at equinox, minimum is about at sunrise; maximum occurs at 2–4 p.m. At June solstice minimum occurs earlier; at December solstice minimum occurs later. Time of maximum value is relatively unchanged.

SOLAR RADIATION AND AIR TEMPERATURES AT HIGH ALTITUDES

SOLAR RADIATION AND AIR TEMPERATURES AT HIGH ALTITUDES 56

Compare the intensity of insolation at high altitudes with that at sea level. Give reasons for the difference.

Solar radiation (insolation) much more intense at high altitude because atmospheric absorption and energy loss by scatter are much less.

How is the daily cycle of air temperature affected by increased altitude?

At higher altitudes, daily air temperature cycle show greater range.

How is the mean daily, monthly, or annual temperature affected by increase in altitude?

Mean daily, monthly, and annual air temperatures decrease with increase in altitude. Otherwise, annual cycle is similar in form. (See also p. 150, and Figure 9.31 for annual temperature cycle.)

DAILY CYCLE OF SOIL TEMPERATURE

How does the daily cycle of temperature at the soil surface compare with that at thermometer shelter level? Explain.

Describe the daily soil temperature cycle with increasing depth in soil.

DAILY CYCLE OF SOIL TEMPERATURE 57

Daily temperature cycle at soil surface shows much greater range than at thermometer shelter level. Peak temperature is much lower. Effect is strongest on bare ground or city pavement; weak under forest cover.

At increasing depth, daily soil temperature cycle shows reduced amplitude; phase lags with increasing depth. Cycle is imperceptible below about 1 m.

TEMPERATURE INVERSION AND FROST

Define *low-level temperature inversion.* How is the normal environmental lapse rate affected? Describe the lapse rate curve when inversion is present.

What is the cause of low-level inversion?

Define *killing frost.* What has frost to do with growing season? How is killing frost related to low-level temperature inversion?

TEMPERATURE INVERSION AND FROST 58

Low-level temperature inversion: reversal of normal environmental lapse rate curve to give temperature increase upward from ground level. Normal temperature decrease sets in above inversion layer.

Cause: heat loss in soil layer and lower air layer by radiational cooling at night.

Killing frost: period of below-freezing air temperatures sufficiently long to kill sensitive crops; ends growing season. May occur during period of low-level temperature inversion.

ANNUAL CYCLE OF AIR TEMPERATURE

Describe the way in which the annual cycle of net radiation is plotted. What values are entered on the graph?

Describe the way in which monthly mean air temperature changes as net radiation rises to a surplus and falls to a deficit.

Describe the annual cycles of net radiation and mean temperature in the equatorial zone. Why is the annual range small?

Define *annual temperature range.*

ANNUAL CYCLE OF AIR TEMPERATURE 58

Net radiation cycle is plotted in terms of monthly mean value of daily net radiation. (1y/day).

Monthly mean air temperature rises when net radiation is an increasing surplus quantity; falls as surplus declines.

Equatorial zone: sustained large energy surplus and uniform warm temperatures. Annual temperature cycle weak; very small range.

Annual temperature range: difference between highest-month mean and lowest-month mean.

ANNUAL CYCLE OF SOIL TEMPERATURE

Describe the changes in annual soil temperature cycle with increasing depth.

ANNUAL CYCLE OF SOIL TEMPERATURE 60

Annual cycle of soil temperature shows decreasing range with greater depth; cycle also shows lag in phase (peak value occurs later in year).

How is constant temperature deep in soil or rock related to air temperature?

Below depth affected by annual cycle temperature of soil or rock is constant at a value slightly higher than mean annual air temperature.

ANNUAL CYCLE OF WATER TEMPERATURE IN LAKES AND OCEANS

ANNUAL CYCLE OF WATER TEMPERATURE IN LAKES AND OCEANS 60

What is the *epilimnion*?

Epilimnion: isothermal, upper warm layer of a lake; uniform temperature.

What is the *hypolimnion?*

Hypolimnion: deep, cold water layer of a lake found below the thermocline; uniform temperature: 4°C (39F).

LAND AND WATER TEMPERATURE CONTRASTS

LAND AND WATER TEMPERATURE CONTRASTS 62

What is the principle behind the differences in heating and cooling rates of water bodies versus land bodies? Which type of body heats and cools more rapidly? which more slowly?

Principle: surface of large water body heats more slowly and cools more slowly than surface of large land body, when both are subject to same intensity of insolation. Always make comparisons at same latitude.

State four physical reasons for the slower heating and cooling of water bodies.

Water body: (1) radiation penetrates surface layer; (2) specific heat large; (3) water layer is mixed; (4) evaporation cools surface.

State four reasons for the more rapid heating and cooling of land bodies.

Land body: (1) soil is opaque; (2) specific heat of soil is small; (3) dry soil is poor conductor; (4) no mixing in soil.

How does the principle of land and water contrasts have an effect upon the daily air temperature curve? Where is range highest?

Effect upon daily air temperature curves: daily range high in interior desert location; low close to ocean.

How do land and water contrasts have an effect upon the annual cycle of air temperature? Where is range greatest? where least? Name the months of maximum and minimum temperatures on lands; on oceans.

Effect upon annual cycle of air temperature: annual range large in continental interior; small at coastal or island location. July and January are maximum and minimum months inland on continents. August and February are maximum and minimum months close to ocean.

AIR TEMPERATURE MAPS

AIR TEMPERATURE MAPS 63

Define *isotherm*. How are isotherms constructed? What do they show?

Isotherm: line drawn on map to connect all points having same temperature reading.

WORLD PATTERNS OF AIR TEMPERATURE

For what months of the year are monthly mean temperatures compared?

Name the three controls of patterns of isotherms.

Describe the effect of latitude upon isotherms. In what part of the world is this effect best shown? Why? Explain the general decrease of temperature with increase in latitude.

Describe the effect of landmasses upon isotherms. Where are the centers of most intense cold at subarctic latitudes?

Describe the annual latitude migration of isotherms. Over what regions is migration small? Where is it very large? Explain.

How do highlands influence air temperatures and isotherms?

What temperature conditions are found over areas of perpetual ice and snow? Name the three principal areas. Why is it coldest over the ice sheets?

WORLD PATTERNS OF AIR TEMPERATURE 63

July and January monthly mean temperatures compared.

Controls of patterns of isotherms: (1) latitude; (2) continent-ocean contrasts; (3) altitude.

Effect of latitude: isotherms trend east-west, temperatures decrease into higher latitudes. Best shown in southern hemisphere. Effect due to decreasing insolation with latitude.

Effect of landmasses: develops cold centers in winter at subarctic latitudes, esp. Siberia and North America.

Annual latitude migration of isotherms: very small over equatorial zone oceans. Very large migration over continents in midlatitude and subarctic zone, esp. in North America and Eurasia.

Highlands always colder than adjacent lowlands.

Areas of perpetual ice and snow always intensely cold: Greenland, Antarctica, Arctic Ocean. Coldest over ice sheets because of high altitude.

THE ANNUAL RANGE OF AIR TEMPERATURE

Where do the greatest annual ranges of air temperature occur?

Where are moderate annual ranges found?

Where is annual range least?

THE ANNUAL RANGE OF AIR TEMPERATURE 66

Range extremely great in subarctic and arctic zones of Asia and North America.

Range moderate over large land areas in tropical zone.

Range very small over oceans in equatorial zone.

WORLD PATTERNS OF SEA SURFACE TEMPERATURE

Describe the general global pattern of sea surface temperatures.

Describe the annual range of sea surface temperatures.

WORLD PATTERNS OF SEA SURFACE TEMPERATURE 67

Sea surface temperatures trend east-west generally around the globe throughout the year, with maximum values in a low-latitude belt and strong poleward decrease both north and south.

Equatorial oceans maintain temperatures above 26°C throughout year; arctic waters remain close to the freezing point, 0°C; middle latitudes show large annual range.

OCEAN THERMAL ENERGY

What is the principle used for extraction of thermal energy from the ocean?

OCEAN THERMAL ENERGY **68**

Uses temperature gradient from surface downward in tropical waters.

GREENHOUSE EFFECT AND GLOBAL CLIMATE CHANGE 68

List three ways in which humans induce atmospheric change.

How humans induce atmospheric change: (1) change in gaseous components of atmosphere; (2) emissions of industrial gases and dust; (3) changes in plant cover and soil surface of lands.

CARBON DIOXIDE ON THE INCREASE

Has carbon dioxide content of the atmosphere increased in the past century? By how much? What is the projected increase?

What is the percentage of CO_2 in atmosphere contributing to greenhouse effect?

CARBON DIOXIDE ON THE INCREASE **68**

Carbon dioxide increase: 22% increase in past 130 years. Projected increase of 35% by year 2000.

Percentage of CO_2 contribution to greenhouse effect is 60%.

THE ROLE OF TRACE GASES IN THE GREENHOUSE EFFECT

What are the *trace gases?*

How do trace gases contribute to greenhouse effect?

What is the combined percentage contribution of CFC to the greehouse effect?

THE ROLE OF TRACE GASES IN THE GREENHOUSE EFFECT **69**

Trace gases include: methane (CH_4), nitrous oxide (N_2O), ozone (O_3), and two types of chlorofluorocarbons (CFC).

Trace gases absorb outgoing longwave radiation throughout the troposphere.

The combined percentage contribution to CFC on the greenhouse effect is 12%.

HAS GLOBAL WARMING SET IN?

What trends are evident in global temperatures since 1856?

What trends do temperatures inferred from tree-ring data show from 1700 to present? How do these temperatures compare with global temperatures since 1856?

What is the "Litttle Ice Age"?

What has been the effect of increased CO_2, methane, ozone, and CFC on global temperatures?

What do scientists predict about future rate of global warming? How does the most extreme prediction compare with rates of change seen in tree-rings or ice-cores?

HAS GLOBAL WARMING SET IN? **70**

Since 1856, global temperature trends show a clear increase, but with wide swings on a time scale of a decade or so.

Temperatures since 1700 inferred from tree-rings show two strong periods of increase with a decrease in between. The last increase fits global temperatures since 1856 well.

The Little Ice Age is a cold period with a low point around 1840.

The effect of CO_2, methane, ozone and CFC's in global temperatures in generally not believed to be measurable yet.

Scientists predict that temperatues will increase by 0.06°C to 0.80C per year, depending on the scenario. The most rapid rate of increase is 10 to 100 times faster than that seen in tree-ring or ice-core data.

CHAPTER 4—SAMPLE OBJECTIVE TEST QUESTIONS

A. MATCHING

1.	thermometer	a. ___1___	temperature map
2.	hypolimnion	b. ___3.___	maximum-minimum
3.	inversion	c. ___4___	climate change
4.	isotherm	d. ______	deep lake layer
5.	carbon dioxide	e. ___2___	killing frost

B. MULTIPLE CHOICE

1. The mean monthly air temperature is calculated as the

   ______a. mean of daily means for the entire year.

   ______b. sum of daily maximum and minimum divided by two.

   ___X___c. mean of daily means for a given month.

   ______d. mean of monthly means for all months of year.

2. In a low-level temperature inversion the air temperature from ground surface upward

   ______a. first increases, then decreases.

   ______b. first decreases, then increases.

   ______c. first holds constant, then decreases.

   ______d. first decreases, then holds constant.

3. One reason that a large water body heats more slowly than a large land body is because

   ______a. specific heat of water is small.

   ______b. evaporation warms the water surface.

   ______c. radiation cannot penetrate the surface water layer.

   ______d. the surface water layer is easily mixed.

C. COMPLETION

1. The warm upper water layer of a lake is called the ______________________________.
2. A period of below-freezing temperature sufficiently long to kill sensitive plants is a ______________________________.
3. The difference between highest-month mean and lowest-month mean air temperatures is the ______________________________.

4. A line drawn upon a map to connect all points having the same air temperature is an ________

_________________________.

5. Coldest air temperatures on an annual average basis are found over _________________________

_________________________ because of their high altitude and high latitude.

6. An atmospheric gas whose quantity is known to have risen steadily in the past century is_____

_________________________.

CHAPTER 5

WINDS AND THE GLOBAL CIRCULATION

PAGE

WINDS AND THE PRESSURE GRADIENT FORCE

Define *wind.* What is the dominant plane of motion in wind?

What is an *isobaric surface?*

Define *isobar.*

What is the *pressure gradient?* In what direction is it measured with respect to isobars?

Explain the *pressure gradient force.* What determines the magnitude of the force? How is the force related to steepness of gradient?

WINDS AND THE PRESSURE GRADIENT FORCE 72

Wind: air motion relative to the earth's surface. Motion is dominantly horizontal.

Isobaric surface: surface of equal barometric pressure.

Isobar: line on map passing through all points having the same barometric pressure.

Pressure gradient: change of pressure measured along a line at right angles to isobars.

Pressure gradient force: force acting horizontally tending to move air in the direction of lower pressure. Force is directly proportional to steepness of pressure gradient (steep gradient = strong force).

SEA AND LAND BREEZES

What is a *sea breeze?* In which direction does it blow? What causes the sea breeze? Explain.

What is a *land breeze?* In which direction does it blow? What causes the land breeze? Explain.

SEA AND LAND BREEZES 73

Sea breeze: local wind blowing from sea to land during day. Heating of coastal ground surface causes lower barometric pressure, with gradient from sea to land.

Land breeze: local wind from land to sea during night. Cooling of coastal ground surface raises barometric pressure, with gradient from land to sea.

MEASUREMENT OF WINDS

What is a *wind vane?* How is a wind vane constructed? What does it tell?

MEASUREMENT OF WINDS 73

Wind vane: pivoted horizontal arm free to swing so as to point into wind. Tells wind direction.

Describe the *anemometer.* What does it measure? Describe the cup-type anemometer. How is it constructed?

Anemometer: any instrument that measures wind speed. Cup anemometer uses cups mounted on vertical shaft; rotation speed proportional to wind speed.

What equipment is used to measure upper-air speeds?

Sounding balloons used to measure upper-air wind speeds.

Measurement of wind direction and speed throughout the troposphere has been facilitated by what system?

The Doppler radar system (NEXRAD) has facilitated measurement of wind direction and speed.

THE CORIOLIS EFFECT AND WINDS

THE CORIOLIS EFFECT AND WINDS 74

What is the *Coriolis effect?* State direction in which the effect acts in the northern hemisphere and in the southern hemisphere. What causes the Coriolis effect? How does the intensity change from equator to poles? How is the intensity related to speed of the object in motion?

Coriolis effect: a fictitious force tending to deflect any object in motion toward the right of its direction of motion in the northern hemisphere; toward the left in the southern hemisphere. Earth rotation causes Coriolis effect. Intensity is zero at equator; increases to maximum at poles. Intensity increases with increasing speed of the object in motion.

Explain how the Coriolis effect acts upon air in motion. What direction of motion is achieved with respect to isobars?

Coriolis effect acts upon air in motion to turn the direction of flow until it is parallel with isobars, where no ground friction is present.

How does friction with the ground affect direction of surface winds? How are surface winds related in direction to isobars?

Surface winds: friction with ground decreases Coriolis effect; wind makes angle with isobars.

State Ballot's law.

Ballot's law for northern hemisphere: stand with your back to the wind and low pressure will be toward your left, high pressure toward your right.

How is wind speed related to the spacing of isobars?

Close spacing of isobars means strong winds; wide spacing means weak winds.

CYCLONES AND ANTICYCLONES

CYCLONES AND ANTICYCLONES 75

Define *cyclone.* Is the pressure in a cyclone center "high" or "low"?

Cyclone: a center of low barometric pressure, or "low."

Define *anticyclone.* Is it a center of "high" or "low" pressure?

Anticyclone: a center of high barometric pressure, or "high."

What pattern do isobars make around cyclones and anticyclones?

Isobars form concentric patterns around cyclones and anticyclones. Make a sketch map (see Figure 5.9, pg. 75).

Describe the direction of motion of surface winds within a cyclone in the northern hemisphere; in the southern hemisphere.

Surface winds spiral anticlockwise inward toward center of cyclone in northern hemisphere; clockwise in southern hemisphere. Make a sketch map (see Figure 5.9, pg. 75).

Describe the direction of motion of surface winds within an anticyclone in the northern hemisphere; in the southern hemisphere.

Surface winds spiral clockwise outward from center of anticyclone in northern hemisphere; anticlockwise in southern hemisphere. Make a sketch map.

Explain how convergence of air occurs in a cyclone. Does the air rise or sink?

Convergence of air occurs in cyclones; requires rise of air in center.

Explain how divergence of air occurs in an anticyclone. Does the air rise or subside?

Divergence of air occurs in anticyclones; requires subsidence of air in center.

GLOBAL DISTRIBUTION OF SURFACE PRESSURE SYSTEMS

GLOBAL DISTRIBUTION OF SURFACE PRESSURE SYSTEMS 75

What do isobaric maps show? How are readings adjusted for station altitude?

Isobaric maps show pressure distribution. Readings are corrected for station altitude to give sea-level values.

Define *equatorial trough.* Where does it lie? What pressures are typical?

Equatorial trough: an east-west trough of weak low pressure, 1011 to 1008 mb, lying approximately over the equator.

Define *subtropical belts of high pressure.* What pressures are found in these belts? Where are they located?

Subtropical belts of high pressure: belts of high pressure, over 1020 mb, centered about on latitudes 30° N and S.

What are *pressure cells?* Where are they located?

Pressure cells: centers of high pressure (anticyclones) within the subtropical belts of high pressure.

Define *subantarctic low-pressure belt.* What are the lowest pressures in this belt? Where is it found?

Subantarctic low-pressure belt: trough of low pressure, down to 984 mb, over Southern Ocean at about 65° S.

What is the *polar high?* Where is it found?

Polar high: high-pressure center (anticyclone) over south pole and antarctic continent.

NORTHERN HEMISPHERE PRESSURE CENTERS

NORTHERN HEMISPHERE PRESSURE CENTERS 78

Describe the way in which North America and Eurasia dominate pressure systems. Explain.

Effect of North America and Eurasia dominates pressure systems. (Land areas cool more rapidly and intensely than ocean areas at the same latitude.)

Describe winter pressure conditions over Eurasia and North America.

Winter: *Siberian high,* pressure over 1030 mb, and *Canadian high* over continents.

What pressure centers exist over the North Pacific and North Atlantic?

Aleutian low and *Icelandic low* over oceans.

Describe summer pressure conditions over southern Asia. Is pressure high or low? Explain. What pressure centers occur over the oceans? Give names of these centers over the Atlantic and Pacific.

Summer: strong low over southern Asia; strong highs over oceans (*Azores high* or *Bermuda high, Hawaiian high*).

GLOBAL PATTERN OF SURFACE WINDS

What are the *trade winds (trades)?* What is their direction? Where are they found? Between what pressure belts?

What is the *intertropical convergence zone* (ITC)? Where is it found? What winds converge in this zone?

What are the *doldrums?* Where do the doldrums occur? With what weather are they associated ?

Describe the seasonal north-south shift of trades, ITC, and doldrums. Where is the shift greatest? where least?

What are the *horse latitudes?* What conditions of pressure and winds prevail in this zone?

What are the *prevailing westerly winds (westerlies)?* Where are these winds located? In what hemisphere are they best developed?

Where do the *polar easterlies* occur? In what part of the world are they well developed?

GLOBAL PATTERN OF SURFACE WINDS 78

Trade winds (trades): persistent easterly winds developed between subtropical highs and equatorial trough.

Intertropical convergence zone (ITC): zone of convergence of trades along axis of equatorial trough.

Doldrums: belt of calms and variable winds occurring at times along the equatorial trough.

Seasonal north-south shift of trades, ITC, and doldrums least over Pacific and Atlantic oceans, greatest over South America, Africa, and Indian Ocean-S.E. Asia region.

Horse latitudes: region of frequent calms near centers of subtropical; high-pressure cells.

Prevailing westerly winds (westerlies): predominantly westerly winds in latitude zone 35° to 60°. Best developed in southern hemisphere.

Polar easterlies: easterly winds in arctic and polar zones. Not a dominant system in northern hemisphere. Easterly winds blow off antarctic ice sheet, out from polar high.

MONSOON WINDS OF SOUTHEAST ASIA

Define *summer monsoon.* In what direction do winds blow? What pressure conditions prevail? What is the associated weather?

Define *winter monsoon.* In what direction do winds blow? What are the pressure conditions? What kind of weather prevails?

MONSOON WINDS OF SOUTHEAST ASIA 80

Summer monsoon: inflow of air from Indian Ocean toward Asiatic low pressure center. Associated with rainfall.

Winter monsoon: inflow of cool, dry air from Siberian high southward over Southeast Asia.

LOCAL WINDS

Define *local winds.* What causes local winds?

Describe *mountain and valley winds.* In what directions does the wind blow? How do wind directions change between night and day? What is the cause of these winds?

LOCAL WINDS 81

Local winds: winds of local extent generated by terrain effects.

Mountain and valley winds: alternating air flow upward over mountain slopes during day and downward into valleys at night. Caused by local heating and cooling of slopes.

What are *drainage winds?* What causes the winds? Give an example.

Drainage winds (katabatic winds): gravity flow of colder, denser air from interior region through mountain passes to adjacent lowlands. Examples: Antarctic blizzard winds, *mistral* of France.

Describe the *Santa Ana* wind. Where does it occur?

Santa Ana: hot, dry easterly wind that blows outward from desert in southern California. Carries dust; fans brush fires.

What is the *bora?* Where does it occur?

Bora: local wind of the Adriatic coast; a strong, cold winter wind produced by a strong pressure gradient.

What are *foehn* or *chinook* winds?

Foehn or *chinook* winds: result when strong regional winds pass over a mountain range and are forced to descend on the lee side. The air is heated and dried.

WINDS ALOFT

WINDS ALOFT 81

How is wind direction related to isobars at high altitude? Explain.

Winds follow isobars at high altitude. Surface friction absent. Coriolis effect turns wind until parallel with isobars.

What is the *geostrophic wind?*

Geostrophic wind: wind at high levels above the earth's surface blowing parallel with a system of straight, parallel isobars. (Does not apply to curved isobars, because a curved path introduces centrifugal force.)

WINDS ON A NONROTATING PLANET

WINDS ON A NONROTATION PLANET 82

Describe the hypothetical wind system for a nonrotating planet.

On imaginary nonrotating planet, air rises in equatorial low, travels poleward at high level, sinks over polar high, flows equatorward at surface, setting up simple two-cell system of *meridional winds.*

Define *meridional winds.*

Meridional winds: winds moving across the parallels of latitude in a north-south or south-north direction along the meridians of longitude.

What is a *heat engine?*

Heat engine: a mechanical system in which motion is powered by inflow of heat energy. System of meridional winds represents a simple heat engine.

THE HADLEY CELL CIRCULATION

THE HADLEY CELL CIRCULATION 82

What happens to poleward-moving air as it reaches the tropical zone?

Poleward-moving air at high levels, upon reaching the tropical zone (lat. 20°–30°) tends to accumulate (pile up) and undergoes subsidence.

Define subsidence.

Subsidence: downsinking of a large mass of air in the central region of an anticyclone (high-pressure region).

What are the *tropical easterlies (equatorial easterlies)?* What is their direction? Between what pressure belts do they occur?

Tropical easterlies: persistent easterly winds encircle globe over equatorial and tropical zones between subtropical highs.

Describe the *Hadley cell.* Describe air flow within the Hadley cell. Where does air rise? Where does it subside? Explain. In what respect is the Hadley cell a heat engine?

Hadley cell: a circulation cell of meridional air flow between equatorial trough in which air rises, and subtropical highs in which air subsides. Circulation represents a heat engine.

THE UPPER-AIR WESTERLIES AND ROSSBY WAVES

THE UPPER-AIR WESTERLIES AND ROSSBY WAVES 82

What are the *upper-air westerlies?* Describe their flow. How are they related to barometric pressure? to pressure gradient?

Upper-air westerlies: persistent westerly winds form a circumpolar vortex around upper-air *polar low.* Pressure gradient is poleward from weak upper-air subtropical high pressure belt.

What is the *polar low?*

Polar low: persistent center of low barometric pressure over high latitudes in the upper atmosphere.

What are *Rossby waves?* Describe these waves. Where do they occur?

Rossby waves: horizontal undulations in flow path of upper-air westerlies. They form in contact zone between cold *polar air* and warm *tropical air;* this contact zone is the *polar front.*

Define *polar front.* Where does it occur?

Polar front: fluctuating line of contact between masses of polar air and tropical air; found in midlatitude and subarctic zones; a zone of frequent severe atmospheric disturbances (wave cyclones).

Describe the development of a Rossby wave.

Rossby wave, forming on polar front, may deepen and form a detached mass of cold air *(cut-off low)* and a detached mass of warm air *(cut-off high).*

In what way do Rossby waves contribute to global heat transport and the heat balance?

Rossby waves lead to *advection* in which heat is transported poleward across the parallels of latitude.

Define the *advection process.*

Advection process: horizontal mixing of cold and warm air in midlatitude cyclones and anticyclones.

POLAR FRONT JET STREAM

POLAR FRONT JET STREAM 85

What is a *jet stream?*

Jet stream: high-speed flow of air in narrow, tubular zones at high altitudes.

Where is wind speed fastest in a jet stream?

Wind speed is highest in the center line, or *core,* of the jet stream.

Describe the *polar front jet stream.* Where does it occur? How is it related to the Rossby wave system?

Polar front jet stream: variety of jet stream found at the level of the tropopause in middle latitudes. Air moves eastward along contact between polar and tropical air in a Rossby wave.

What causes the strong wind in the polar front jet stream core?

Steep poleward pressure gradient in polar front zone causes strong wind of jet core.

What is the *subtropical jet stream?* Where is is found?

Subtropical jet stream: jet stream of eastward air flow at tropopause level above the Hadley cell, mostly between lat. 20 and 35.

What is the *tropical easterly jet stream?* Where is it found?

Tropical easterly jet stream: jet stream in which air moves westward in tropical zone at very high altitude. Occurs only in northern hemisphere in high-sun season.

GREENHOUSE WARMING AND THE GLOBAL CIRCULATION

GREENHOUSE WARMING AND THE GLOBAL CIRCULATION 86

What is the primary effect of fossil fuel burning in global climate? Is sensible heat release important?

The primary effect of fossil fuel burning is to increase the amount of greenhouse gases in the atmosphere. Sensible heat release is too small, compared to incoming solar energy, to be important.

What effects can be anticipated from elevation of the mean global air temperature?

With an elevated level of global air temperature, Hadley cell circulation will be strengthened. Subtropical deserts will shift poleward. Intensification of easterly trades and monsoon winds may occur. The polar jet may become more powerful, bringing more frequent and more intense storms.

OCEAN CURRENTS

OCEAN CURRENTS 87

Define *ocean current.* What is the principal cause of ocean currents? Does the Coriolis effect influence currents? Explain.

Ocean current: persistent, dominantly horizontal flow of ocean water. Surface currents largely produced by prevailing winds. Coriolis effect deflects current motion.

What other causes exist for ocean currents?

Differences in water density may also set ocean currents in motion. Chilled surface water sinks.

GENERALIZED SCHEME OF OCEAN CURRENTS

GENERALIZED SCHEME OF OCEAN CURRENTS 87

What are *gyres?* Where are the gyres located? Describe the motion within gyres.

Gyres: large circular current systems centered upon oceanic subtropical highs. Clockwise in northern hemisphere.

What is the *equatorial current?* Where is it? What is the direction?

Equatorial current: west-flowing current in belt of trades.

What is the *west-wind drift?* In what wind belt does it occur? What is the direction of flow?

West-wind drift: slow eastward drift of surface water in belt of westerly winds in midlatitude zone.

What is the *equatorial countercurrent?*

The *equatorial countercurrent:* narrow ocean current, flowing west to east between the two equatorial currents.

Where are warm ocean currents located? In what direction do they move? How are they related to gyres? Give and example.

Warm currents are on west sides of oceans; they move poleward in gyres. Examples: Gulf Stream, Japan current.

Where are cool currents found? In what direction do they move? How does upwelling influence these currents. Give an example.

Cool currents on east sides of oceans. Move equatorward in gyres. Upwelling of cold water contributes to coolness. Examples: Humboldt current, Benguela current.

Where are cold currents found? What are cold gyres? Where are they?

Cold currents issue from Arctic Ocean into North Atlantic and North Pacific. Cold gyres circulate around Icelandic and Aleutian lows.

What is the *antarctic circumpolar current?* In which direction does it move?

Antarctic circumpolar current: east-moving cold current in continuous ocean belt surrounding continent of Antarctica.

OCEANIC STREAMS AND THEIR EDDIES

OCEANIC STREAMS AND THEIR EDDIES 88

In which direction is the *warm-core ring* rotating?

The *warm-core ring* is rotating in a clockwise direction.

In which direction is the *cold-core ring* rotating?

The *cold-core ring* is rotating in an anticlockwise direction.

What is the process by which warm-core and cold-core rings are formed?

Warm-core and cold-core rings are formed by the cutoff of eddies in an ocean current, such as the Gulf Stream.

What is the process by which warm-core and cold-core rings are found?

Warm-core and cold-core rings are found by the cutoff of eddies in an ocean current, such as the Gulf Stream.

WIND AS AN ENERGY RESOURCE 90

Evaluate wind power as an energy resource.

Global supply of wind energy at favorable sites estimated to be 20 million megawatts.

How is wind power developed?

Wind power makes use of windmills, which transform kinetic energy of mechanical motion into electricity.

What is a windfarm?

A windfarm is a collection of wind turbines at a suitable location that is used to generate electric power.

What regions and countries have led in the development of wind power?

Regions and countries that have led in the development of wind power include California, Denmark, Holland, and the United Kingdom.

WAVE POWER

WAVE POWER 92

How can wave energy be utilized?

Wave energy in kinetic form in wave orbits can be transferred to rising and falling tethered floating objects, operating electrical generator mechanism.

CURRENT POWER

How might an ocean current be developed as an energy resource?

CURRENT POWER 92

Narrow, swift ocean currents, such as Gulf Stream, might be used to turn turbines tethered to ocean floor.

CHAPTER 5 —SAMPLE OBJECTIVE TEST QUESTIONS

A. MATCHING

1. sea breeze
2. wind vane
3. wind speed
4. Coriolis effect
5. cyclone
6. equatorial trough
7. trade winds
8. doldrums
9. Rossby waves
10. gyres

a. ________ calms
b. ________ ocean currents
c. ________ earth rotation
d. ________ low pressure
e. ________ jet stream
f. ________ heat engine
g. ________ anemometer
h. ________ wind direction
i. ________ easterly winds
j. ________ inward spiral

B. MULTIPLE CHOICE

1. The pressure gradient force

   ________ a. acts parallel with isobars.

   ________ b. acts at right angles to isobars.

   ________ c. is in a direction from low to high pressure.

   ________ d. is opposite to the Coriolis effect.

2. During the summer monsoon in Southeast Asia

   ________ a. weather is usually clear and dry.

   ________ b. winds blow southward from central Asia.

   ________ c. air flows into Asia from the Indian Ocean.

   ________ d. air flows from low to high pressure.

3. Gyres of ocean currents

   ________ a. turn clockwise in the northern hemisphere.

   ________ b. turn anticlockwise in the northern hemisphere.

   ________ c. flow opposite in direction to the west-wind drift.

   ________ d. connect the Arctic Ocean with the North Atlantic.

C. COMPLETION

1. A surface of equal barometric pressure is a ________________________.
2. A local wind blowing from sea to land during the day is a ________________________.

3. A circulation cell between the equatorial trough and subtropical high is the ________________

__________________________.

4. High-speed flow of air in narrow tubular zones at high altitudes is known as a ______________

__________________________.

5. The slow, eastward drift of surface water in the belt of westerly winds is the ________________

__________________________.

CHAPTER 6

ATMOSPHERIC MOISTURE AND PRECIPITATION

PAGE

WATER STATES AND HEAT — 93

Name and define the three states in which water occurs.

Water occurs in three states: (1) solid state, frozen as crystalline ice; (2) liquid state, as water; (3) gaseous state, as water vapor.

What processes are involved in changes of state?

Changes of state involve *condensation, sublimation, evaporation, freezing,* and *melting.*

Define *condensation.*

Condensation: process of change of matter from gaseous state (water vapor) to liquid state (liquid water) or solid state (ice).

Define *sublimation.*

Sublimation: process of change of water vapor (gaseous state) directly to ice (solid state), or vice versa.

Define *evaporation.*

Evaporation: process in which water in the liquid state or solid state passes into the vapor state.

Define *freezing.*

Freezing: change of state of water from liquid state to solid state.

Define *melting.*

Melting: change of state of water from solid state to liquid state.

What exchange of energy occurs during evaporation or sublimation?

During evaporation or sublimation, sensible heat is transformed into *latent heat of vaporization* and held in the water vapor.

What is *latent heat?*

Latent heat: form of heat absorbed and held in storage in a gas or a liquid during the processes of evaporation or melting, respectively; distinguished from sensible heat.

How much *latent heat of vaporization* is absorbed during the evaporation of a gram of water?

Latent heat of vaporization, absorbed by water vapor during evaporation, is about 600 cal per gram of water evaporated.

Can latent heat be liberated? Explain.

During condensation of water vapor, the latent heat it holds is liberated to become sensible heat; this process raises (or tends to raise) the sensible temperature of the surrounding air.

How much heat is liberated or absorbed during melting or freezing?

Latent heat of fusion, about 80 cal per gram of water, is liberated during melting or absorbed during freezing.

HUMIDITY

HUMIDITY 93

Define *humidity.*

Humidity: amount of water vapor present in the air.

What is *saturated air?* Why must temperature be specified?

Saturated air: air holding the maximum quantity of water vapor possible at a specified temperature. Quantity increases with temperature.

Define *relative humidity.* In what units is relative humidity stated?

Relative humidity: ratio or water vapor present to maximum quantity at saturation, expressed as a percentage.

In what two ways can the relative humidity be increased within a given body of air?

Increase in relative humidity by (1) evaporation and diffusion from free water surface; (2) decrease in air temperature.

Describe the typical daily cycle of relative humidity. Explain the daily rise in humidity; the daily fall in humidity.

Daily cycle of relative humidity; decrease during day as air temperature rises; increase at night as air temperature falls.

Define *dew point.* In what units is it expressed? What happens when air temperature is reduced below the dew point?

Dew point: critical air temperature at which air is saturated. Temperature reduction below dew point normally results in condensation as dew or frost.

HOW RELATIVE HUMIDITY IS MEASURED

HOW RELATIVE HUMIDITY IS MEASURED 94

Describe a simple *hygrometer.*

Hygrometer: instrument that measures relative humidity. May use strand of human hair.

What is a *hygrograph?*

Hygrograph: instrument that continuously records relative humidity.

Describe the *sling psychrometer.* How does it work?

Sling psychrometer: pair of thermometers, one with dry bulb and one with wet bulb, moved rapidly through air. Degree of cooling of wet bulb by evaporation is greater for air of lower humidity.

SPECIFIC HUMIDITY

SPECIFIC HUMIDITY 95

Define *specific humidity.* What units are used? Give typical value of specific humidity for cold arctic air; for warm equatorial air.

Specific humidity: mass of water vapor contained in a unit mass of air. Units: gm/kg. Ranges from 0.2 gm/kg in cold arctic air to 18 gm/kg in warm equatorial air.

In what respect does specific humidity describe the capability of air to yield precipitation? Explain.

Specific humidity describes capability of air to yield precipitation.

CONDENSATION AND THE ADIABATIC PROCESS

CONDENSATION AND THE ADIABATIC PROCESS 96

Define *precipitation*. What states of water are involved?

Precipitation: particles of water, liquid or solid, that fall from atmosphere and reach ground.

What must happen to a large mass of air to produce precipitation?

Precipitation requires lift of large mass of air to higher altitude.

What is the *adiabatic process?* What causes change of air temperature in this process? Is heat lost or gained from outside? What effect has compression? expansion?

Adiabatic process: change of temperature within a gas because of compression or expansion, without gain or loss of heat from outside. Compression causes heating; expansion causes cooling.

What change in volume and temperature does rising air experience?

Rising air expands, undergoes adiabatic cooling.

Define *dry adiabatic lapse rate*. Give the value of this lapse rate. How does it differ from the environmental temperature lapse rate?

Dry adiabatic lapse rate: rate at which air is cooled as it rises, when no condensation is occurring. Value: 10 C°/1000 m (5.5 F°/1000 ft). (Do not confuse with environmental temperature lapse rate; see Chapter 2, pg. 25.)

How does the dew point change within a mass of air as it rises? Give the rate of change.

Dew-point temperature also decreases with increasing altitude. Rate: 2 C°/1000 m (1 F°/1000 ft).

At what point in a rising mass of air does condensation begin? How is this point related to the dew point temperature?

Condensation sets in at altitude where adiabatic cooling brings air to dew-point temperature (lines intersect on graph, Figure 6.9, pg. 96).

Define *wet adiabatic lapse rate*. Give the value of this lapse rate. Is it more or less than the dry adiabatic lapse rate? Explain fully.

Wet adiabatic lapse rate: reduced lapse rate when condensation is taking place in rising air. Liberation of latent heat partially offsets adiabatic cooling. Value: 3 to 6 C°/1000 m (2 to 3 F°/1000 ft); varies with condensation rate.

How does the wet adiabatic lapse rate change with increasing altitude? Explain.

As air mass rises, the rate of condensation gradually decreases, approaching zero at high levels; thus wet adiabatic rate declines, approaching dry rate.

CLOUD PARTICLES

CLOUD PARTICLES 97

Define *cloud*. Of what states of water can cloud particles be formed? What is the diameter range of cloud particles?

Cloud: dense mass of suspended water or ice particles in diameter range 20 to 50 microns.

Define *nucleus*. What substance is a common type of nucleus?

Nucleus: core of solid matter upon which water condenses to form cloud particles. Sea salts are common type of nuclei; they are *hygroscopic*.

What is meant by *hygroscopic?*

Hygroscopic: property of a substance to absorb water molecules from the surrounding air.

What is *supercooled* water? At what temperatures do supercooled water droplets occur in clouds? At what temperature are cloud particles entirely ice crystals?

Supercooled water: water in liquid state in droplets well below freezing temperature of 0°C (32°F). Commonly found in clouds. Droplets exist down to about –12°C (10°F). Mixture of water and ice: –12° to –30°C (10° to –20°F). Ice crystals below –30°C (–20°F).

CLOUD FORMS

CLOUD FORMS 97

Define *stratiform clouds.*

Stratiform clouds: layered clouds.

Define *cumuliform clouds.*

Cumuliform clouds: globular clouds.

On what basis are clouds classified?

Clouds are classified on basis of altitude and form. Four *cloud families* are recognized: (1) high; (2) middle; (3) low; (4) clouds of vertical development.

What clouds are found in the high cloud family? Describe *cirrus, cirrostratus,* and *cirrocumulus.*

Family A: high clouds, 6–12 km (20,000–40,000 ft); formed of ice particles.

Cirrus: delicate, wispy, fibrous forms; "mares' tails."

Cirrostratus: veil-like layer cloud, makes halo around sun or moon.

Cirrocumulus: globular cloud masses in rows or layers; "mackerel sky."

Describe the clouds of the middle cloud family, including *altostratus* and *altocumulus.*

Family B: middle clouds, 2–6 km (6,500–20,000 ft).

Altostratus: blanket layer.

Altocumulus: globular mases in layer.

Describe the clouds of the low cloud family, including *stratus, nimbostratus,* and *stratocumulus.*

Family C: low clouds. Ground level to 2 km (0–6,500 ft).

Stratus: dense, low layer.

Nimbostratus: stratus yeilding rain.

Stratocumulus: low layer of globular or roll-like cloud masses.

Describe the cloud family of vertical development, including *cumulus* and *cumulonimbus.*

Family D: clouds with vertical development; extend through great altitude range.

Cumulus: white woolpack masses, associated with fair weather. Dense, congested forms yield showers.

Cumulonimbus: large, high, dense cloud mass, a thunderstorm cloud, yielding rain, hail, thunder, lightning.

FOG

FOG 99

What is *fog?* Is fog a form of cloud?

Fog: cloud layer in contact with land or sea surface, or very close to that surface.

Define *radiation fog.* What causes this type of fog? How is it related to the low-level temperature inversion? Explain. When do radiation fogs most commonly occur?

Radiation fog: fog produced by radiational cooling of basal air layer. Associated with low-level air temperature inversion. Usually nocturnal.

Define *advection fog.* What is the relationship between air layer and ground surface in this kind of fog? Explain.

Advection fog: condensation within moist basal air layer moving over a cold land or water surface. Air loses heat to surface beneath.

PRECIPITATION FORMS

PRECIPITATION FORMS 100

Name the forms of precipitation.

Forms of precipitation: *rain, snow, sleet, hail, glaze.*

Define *rain.* What is the size range of falling rain drops? How do raindrops form in warm clouds and in cold clouds?

Rain: falling water drops, usually 1000 to 2000 microns diameter. Form by coalescence of cloud droplets in warm clouds; by melting of falling snowflakes in cold clouds.

Define *snow.*

Snow: falling ice crystals, clotted together into snowflakes.

Define *sleet.* What comprises particles of sleet? How is sleet formed? What conditions aloft does it indicate?

Sleet: ice grains or pellets produced by freezing of falling rain. Generated in cold air layer below warm layer aloft.

Define *hail.* Of what water state is hail composed? How do hailstones form? What conditions do they indicate aloft?

Hail: large pellets or spheres of ice formed in updrafts of cumulonimbus cloud.

Define *glaze?* Where is glaze formed? How is it formed?

Glaze: ice coating on ground, trees, or wires, formed as rain freezes upon landing.

Define *ice storm.*

Ice storm: occurrence of heavy glaze.

HOW PRECIPITATION IS MEASURED

HOW PRECIPITATION IS MEASURED 100

What units are used in stating amount of precipitation?

Precipitation units: depth of fall per unit of time, e.g., cm/hr, cm/day.

What is a *rain gauge?* Describe the construction of the standard rain gauge. Why is a funnel used?

Rain gauge: device to measure rainfall. Usually uses funnel to narrow water column for ease in measurement.

How is snowfall measured?

Snowfall measured by conversion into water equivalent.

HOW PRECIPITATION IS PRODUCED

HOW PRECIPITATION IS PRODUCED 101

What are the two mechanisms of rise of large masses of air?

Two mechanisms of rise of large masses of air: (1) spontaneous rise; (2) forced rise.

Define *convection.* Is convection spontaneous?

Convection: spontaneous rise of moist air in a convection cell.

Define *convection cell.* Describe a convection cell. Why does the air rise? How can small convection systems be generated?

Convection cell: updraft system in which bubblelike mass of air rises rapidly because of its lower density. Heating of lower air during day can generate small convection systems.

What is the heat source for all strong convection cells? Explain how latent heat provides the energy for convection.

Liberation of latent heat during condensation is energy source for all strong convection cells rising high in atmosphere.

Define *unstable air.* What conditions of moisture and heat are present in unstable air? In what global regions and in what seasons is unstable air most likely to be found?

Unstable air: air sufficiently moist and warm to yield large amounts of heat through condensation, and thus to give rise to convectional activity. Typical of equatorial and tropical zones, and midlatitudes in summer.

How does the environmental lapse rate of an unstable mass of air differ from that of a mass of stable air?

Unstable air mass has steep environmental lapse rate (12C°/1000m, in Figure 6.17, pg. 102). This rate is steeper than the dry adiabatic lapse rate.

Define *stable air.*

Stable air: air having environmental temperature lapse rate less than the dry adiabatic lapse rate; the air resists being lifted.

How can a steep environmental lapse rate occur?

Steep environmental lapse rate develops by heating of lower air from underlying warm ocean or land surface. Often associated with large water vapor content (high specific humidity).

THUNDERSTORMS

THUNDERSTORMS 102

Describe a *thunderstorm.* What cloud form is present? Is convection an important process in the thunderstorm? Describe precipitation in a thunderstorm.

Thunderstorm: an intense, local convectional storm associated with cumulonimbus cloud and yielding heavy precipitation.

What forms of sound and light are emitted by a thunderstorm?

Thunder and lightning, hail, commonly present.

What is a thunderstorm cell? How does the air rise in a cell?

Single thunderstorm consists of a cell in which rising air forms a succession of bubblelike air bodies.

What is the process of *entrainment?*

Entrainment: drawing in of surrounding air by air rising rapidly in a thunderstorm cell.

What form has the top of a thunderstorm?

Flattened *anvil top* of cumulonimbus cloud is formed by downwind motion of ice cloud where prevailing wind is strong.

What is *cloud seeding?* What kind of particles are involved? How does cloud seeding occur in a thunderstorm?

Cloud seeding: fall of ice crystals from anvil top of cumulonimbus cloud, serving as nuclei of condensation at lower levels.

What form of local wind is associated with a thunderstorm?

Strong downdraft strikes ground as destructive squall wind.

MICROBURST! A THREAT TO AIRLINE PASSENGERS

Define *microburst.*

What is the best defense to detect a microburst?

MICROBURST! A THREAT TO AIRLINE PASSENGERS 102

Microburst: a brief onset of intense winds close to the ground beneath a thunderstorm cell.

The best defense to detect a microburst threat is the Doppler radar instrument.

CLOUD SEEDING

How has artificial cloud seeding been carried out in thunderstorms? What seeding material is used? What is the effect?

CLOUD SEEDING 104

Artificial cloud seeding: an attempt to induce convectional precipitation by the introduction of small particles of silver oxide smoke into dense cumulus clouds. Effect is debated by meteorologists.

HAIL AND LIGHTNING—ENVIRONMENTAL HAZARDS

How do hailstones cause damage? What parts of the United States are most susceptible to hail damage? What crops are damaged?

In what way is lightning an environmental hazard?

HAIL AND LIGHTNING—ENVIRONMENTAL HAZARDS 104

Hailstones cause heavy crop losses. Plains and central region of United States lose wheat and corn crops.

Lightning, an electric arc generated in cumulonimbus cloud; a cause of death to humans and livestock, and of fires.

OROGRAPHIC PRECIPITATION

Define orographic.

Define *orographic precipitation.* What causes precipitation to occur? Is the rise of air forced or spontaneous?

Define *rainshadow.* Where does a rainshadow lie with respect to a mountain barrier? What causes aridity? Give an example.

Define *chinook winds.* Where do they occur? (See Chapter 5, pg. 81.)

What is the *foehn?* (See Chapter 5, pg. 81.)

What are *isohyets?*

Give a good illustration of the orographic control of precipitation.

OROGRAPHIC PRECIPITATION 105

Orographic: adjective meaning "pertaining to mountains."

Orographic precipitation: precipitation induced by forced rise of moist air over mountain barrier. Cooling by adiabatic process causes condensation, clouds, precipitation.

Rainshadow: zone of aridity to lee of mountain barrier; result of adiabatic warming of descending air. Example: desert of eastern California, Nevada.

Chinook winds: strong downdrafts in lee of mountain range.

Foehn: wind similar to chinook; occurs on N side of Austrian Alps.

Isohyet: line drawn on map through all points having equal precipitation.

Orographic effect well illustrated by California: high precipitation on windward slopes of coastal ranges and Sierra Nevada; rainshadows in Owens Valley, Death Valley.

LATENT HEAT AND THE GLOBAL BALANCES OF ENERGY AND WATER

Explain how the energy surplus of low latitudes is carried poleward as water vapor. How is the energy released?

Describe the average meridional transport of water vapor.

LATENT HEAT AND THE GLOBAL BALANCES OF ENERGY AND WATER 106

Energy surplus of low latitude zones carried poleward partly in form of latent heat in water vapor. Condensation in midlatitude and high latitude zones releases latent heat.

Meridional transport of water vapor is equatorward in latitude zones 0°–20° N and S in Hadley cell circulation. Poleward transport is most rapid in latitude zones 30°–50 N and S by advection in wave cyclones.

CONVECTIVE STORM AS FLOW SYSTEM OF ENERGY AND MATTER

How can a convective storm be modeled as an open energy system? In what forms are the energy inputs?

How does energy output take place?

How can a convective storm be modeled as an open material flow system?

How can similar models represent the entire earth's atmosphere?

CONVECTIVE STORM AS FLOW SYSTEM OF ENERGY AND MATTER 107

As an open energy system, a convective storm has two forms of energy input: (1) direct solar radiation, raising level of sensible heat in storage; (2) import of latent heat by evaporation of water from ground or water surface.

Energy output from system occurs as longwave radiation, as mass transport of sensible heat in falling precipitation, and by conduction to ground.

As an open material flow system a convective storm has input of water vapor, change of state by condensation, storage as liquid, return to vapor state and storage, and output as liquid (or solid) water.

With minor changes the same models can be applied to atmosphere as an open energy system and as an open material system (subsystem) within the hydrologic cycle.

CLOUD COVER, PRECIPITATION AND GLOBAL WARMING

What is the effect of a rise in global sea surface temperature on evaporation?

What is the effect of increased water vapor on global warming?

What is the effect of increased cloud cover on global warming? Why?

CLOUD COVER, PRECIPITATION AND GLOBAL WARMING 108

A rise in global sea surface temperatures will increase the rate of evaporation.

Water vapor will increase global warming since it is a greenhouse gas.

Increased cloud cover is thought to have a cooling effect, because more solar energy is lost to space through reflection from white cloud tops than is saved through the enhancement of the greenhouse effect.

What will be the effect of an increase in global precipitation?

If global precipitation increases, rainfall in the equatorial zone may increase, as may seasonal rainfall in subtropical belts and the monsoon regions. Tropical deserts and mid-latitude rainshadow deserts may intensify and spread poleward.

AIR POLLUTION AND ITS EFFECTS 109

Define *pollutants.*

Pollutants: foreign matter injected by humans into lower atmosphere as particulate matter and as chemical pollutants.

Define *particulate matter.*

Particulate matter: solid and liquid particles capable of being suspended for long periods in the air.

Define *chemical pollutants.* In what state are these pollutants?

Chemical pollutants: gases other than the normal gaseous constituents of air remote from populated, industrialized regions.

Name five common chemical pollutants, or groups of pollutants. Give the chemical formula for each.

Common chemical pollutants: carbon monoxide (CO), sulfur dioxide (SO_2), oxides of nitrogen (NO, NO_2, NO_3), hydrocarbon compounds, ozone (O_3).

Define *smog.* What are the components of smog?

Smog: mixture of particulate matter and chemical atmospheric condition in stagnant air layers.

Define *haze.*

Haze: atmospheric condition when suspended matter concentrations are less dense obscuring only distant objects.

What are the major sources of pollutants in urban areas?

Pollutants introduced in urban areas, largely by fuel combustion, and in rural areas by smelting, mining, quarrying, farming.

What natural pollutants are found in the atmosphere? Name seven.

Natural pollutants include volcanic dusts, sea salts, pollen, terpenes, smoke of grass and forest fires, blowing dust, bacteria, viruses.

Name the major pollutant sources from human activities.

Major pollutant sources:

What are the principal pollutants from vehicular exhausts?

Vehicular exhausts (CO, hydrocarbons, nitrogen oxides).

What pollutants are produced in electricity generating plants?

Electricity generating plants (SO_2, particulates, fly ash).

How important are space heating and refuse burning as sources?

Space heating, refuse burning are minor sources.

Define *fallout.*

Fallout: gravity fall or particulates, reaching ground.

Define *washout.*

Washout: downsweeping of particulates by precipitation.

Define *photochemical reactions.* Name two pollutants produced by photochemical reactions.

Photochemical reactions: action of sunlight upon pollutant gases to synthesize new toxic compounds or gases. Sulfuric acid derived from SO_2; ozone produced by action of sunlight upon nitrogen oxides and hydrocarbons.

LOW-LEVEL INVERSIONS

LOW-LEVEL INVERSIONS 110

What effect has a steepened environmental lapse rate on upward movement of pollutants over an urban area?

Steepened environmental lapse rate over urban area allows warm air and pollutants to rise until level of stability is reached.

Under what conditions does a low-level temperature inversion occur?

Low-level temperature inversion occurs by air cooling at night; creates very stable situation. Pollutants are trapped beneath the *inversion lid.*

Define *inversion lid.*

Inversion lid: top surface of inversion layer, resisting mixture of colder air below with warmer air above.

UPPER-LEVEL INVERSIONS

UPPER-LEVEL INVERSIONS 111

What causes an *upper-level inversion?*

Upper-level inversion: caused by subsidence of air in anticyclone over basal cool air layer, creating an inversion lid.

Give an example of persistent upper-level inversion.

Los Angeles Basin is an example of upper-air inversion caused by subsidence of dry air over a cool, foggy marine air layer. Smog accumulates in marine air layer.

MODIFICATION OF URBAN CLIMATE

MODIFICATION OF URBAN CLIMATE 112

How does urbanization affect the energy balance? How is transpiration changed? What effect have pavements?

Effects of urbanization on energy balance: reduced transpiration. Pavements and masonry hold more heat than soil. Thermal effect is to convert city to desert environment.

What is a *heat island?* Where is the heat island found?

Heat island: a persistent region of higher air temperatures centered over a city.

What is a *pollution dome?*

Pollution dome: broad, low, dome-shaped layer of polluted air formed over urban area at times when winds are weak or calm.

What is a *pollution plume?*

Pollution plume: trail of polluted air carried downwind from a pollution source.

List other climatic effects of urban air pollution.

Other climatic effects of urban air pollution: (1) reduces visibility and illumination; (2) reduces ultraviolet penetration; (3) increases incidence of fog; (4) increases convectional precipitation.

SOME ENVIRONMENTAL EFFECTS OF ATMOSPHERIC POLLUTION

SOME ENVIRONMENTAL EFFECTS OF ATMOSPHERIC POLLUTION 114

What urban air pollutants are harmful to humans?

Urban air pollutants harmful to humans: sulfuric acid produced from SO_2, ethylene produced from hydrocarbon compounds, ozone, carbon monoxide from vehicular exhausts.

How are plants adversely affected by urban air pollutants?	Plants adversely affected by ozone and SO_2. Crops and trees damaged.
Describe the corrosive action of air pollutants.	Corrosive action of sulfuric acid is severe under polluted urban air; affects masonry metals. Ozone damages exposed rubber.
What metal in polluted air is harmful?	Lead fallout from polluted urban air is health threat. Lead additives to gasoline are being phased out.

ACID DEPOSITION AND ITS EFFECTS 114

What is *acid raid?* What acids are present?	*Acid rain:* rainwater having an abnormally high content of acid ions, principally the sulfate ion of sulfuric acid, and also ions of nitric acid.
What values of pH are observed in acid rain?	Rainwater pH of acid rain lies between 3 and 5, and locally as low as 2.1. Values less than 4 are commonly observed over heavily industrialized areas in middle latitudes.
What is the cause of acid rain?	Acid rain results from input of pollutants of fossil fuel combustion.
What are some harmful effects of acid rain?	Acid rain leads to excessive leaching of plant nutrients from soil and to serious disturbances of aquatic ecosystems of lakes and streams. Numerous lakes in affected areas now have no fish.
Describe *acid deposition.*	*Acid deposition*: fallout of both liquid and solid pollutants.
Name four undersirable environmental effects of acid deposition.	Four undesirable effects of acid deposition are: (1) acidification of lakes and streams; (1) excessive leaching of nutrients from plant foliage and soil; (3) metabolic disturbances to organisms; (4) upsetting balance of predators and prey in aquatic ecosystems.

CHAPTER 6—SAMPLE OBJECTIVE TEST QUESTIONS

A. MATCHING

1.	relative humidity	a. ________	ice storm
2.	specific humidity	b. ________	updraft
3.	adiabatic process	c. ________	hygrometer
4.	stratus	d. ________	mountains
5.	fog	e. ________	gm/kg
6.	snow	f. ________	pollutants
7.	glaze	g. ________	compression
8.	convection	h. ________	ice crystals
9.	orographic	i. ________	advection
10.	acid deposition	j. ________	blanketlike

B. MULTIPLE CHOICE

1. The wet adiabatic lapse rate

________a. is a higher value than the dry adiabatic lapse rate.

________b. involves liberation of latent heat.

________c. is the sum of the dry adiabatic rate and dew-point rate.

________d. indicates that clear dry air is present.

2. In the case of an advection fog

________a. a moist air layer moves over a cold surface.

________b. a cold air layer moves over a warm surface.

________c. warm air lies beneath cold air.

________d. radiation is responsible for the occurrence of fog.

3. Which is <u>not</u> known as a common chemical pollutant

________a carbon monoxide.

________b. nitrous oxide.

________c. carbon dioxide.

________d. ozone.

C. COMPLETION

1. The mass of water vapor contained in a unit mass of air is the ______________________________ rate.

2. The rate at which air is cooled as it rises when no condensation is occurring is the__________ ________________________lapse rate.

3. Fog produced by spontaneous cooling of a basal air layer during the night is known as a ________________________fog.

4. Strong downdrafts on the lee side of a mountain range are ______________________________ winds.

5. A mixture of particulate matter and chemical pollutants over an urban area is called________ ______________________.

6. Downsweeping of particulates by precipitation is ______________________________________ ______________________.

7. Action of sunlight upon pollutant gases to synthesize new toxic compounds is called________ ______________________reaction.

CHAPTER 7

AIR MASSES AND CYCLONIC STORMS

PAGE

TRAVELING CYCLONES

TRAVELING CYCLONES 117

Define *cyclonic storm.* What weather phenomena accompany a cyclonic storm?

Cyclonic storm: intense weather disturbance within a moving cyclone generating strong winds, cloudiness, and precipitation.

Name the three classes of moving cyclones.

Three classes of moving cyclones: (1) wave, cyclone; (2) tropical cyclone; (3) tornado.

AIR MASSES

AIR MASSES 117

Define *air mass.* What properties does an air mass have? How high does an air mass extend? What are the horizontal dimensions of an air mass?

Air mass: extensive body of air within which upward gradients of temperature and moisture are fairly uniform. Extends through troposphere; is of subcontinental dimensions.

Define *front.*

Front: surface of contact between two unlike air masses.

Define *source regions.* In what way are source regions classified? What determines the thermal properties of the air mass? What determines the moisture content?

Source regions: extensive land or ocean surfaces over which air masses derive their temperature and moisture characteristics. Source regions classified by: (1) latitudinal position, determining thermal properties, and; (2) underlying surface, whether land or ocean, determining moisture content.

Name five air mass types classified according to latitude. Give capital letter code for each.

Air mass types, according to latitude (capital letter): Arctic (A), Antarctic (AA), Polar (P), Tropical (T), Equatorial (E).

What are the two types of air masses classified according to underlying surface? What code letters are used?

Air mass types, according to underlying surface (lowercase letter): maritime (m), continental (c).

For each of the air masses listed below, give letter code, source region, temperature, and moisture content:

Air masses recognized by combining letters:

Continental arctic (antarctic)

Continental arctic (cA), and continental antarctic (cAA), both very cold, very dry.

Continental polar

Continental polar (cP), cold, dry.

Maritime polar

Maritime polar (mP), cool, moist.

Continental tropical

Continental tropical (cT), warm dry.

Maritime tropical

Maritime tropical (mT), warm, very moist.

Maritime equatorial

Maritime equatorial (mE), very warm, very moist.

NORTH AMERICAN AIR MASSES

Name the typical North American air masses; give source region and character for each.

NORTH AMERICAN AIR MASSES 119

North American air masses:

Air mass	Source Region	Character
cP	North-central Canada	Cold, dry
cA	Arctic Ocean	Very cold dry
mP	North Pacific, North Atlantic	Cool, moist
mT	Gulf of Mexico, Atlantic Ocean, Pacific Ocean	Warm, moist
cT	Northern Mexico, SW United States	Hot, dry

COLD AND WARM FRONTS

What is a *cold front?* Describe the positions of cold and warm air. How steep is the front? What weather conditions usually accompany a cold front?

What is a *warm front?* How are warm and cold air masses related in a warm front? How step is the front? What weather conditions and clouds accompany a warm front?

Describe an *occluded front.* What fronts meet to form an occluded front? Describe conditions aloft.

COLD AND WARM FRONTS 119

Cold front: front along which cold air mass is pushing beneath a warm air mass. Cold air replaces warm at surface. Slope, steep. Usually associated with turbulence, convectional activity, and cumulonimbus clouds.

Warm front: front along which warm air mass is forced to rise over basal cold air mass. Warm air replaces cold at surface. Slope gentle. Often stable atmospheric conditions with stratiform clouds.

Occluded front: composite form of two fronts in which cold front has overtaken warm front, forcing warm air aloft.

WAVE CYCLONES

What is a *wave cyclone?* How are fronts involved in the cyclone?

What is the *wave theory?* Who developed the concept? When?

Describe the stages in the life history of a wave cyclone. What conditions are present prior to wave formation? Where is the center of low pressure situated? What form do the fronts take?

How does the form of the cyclone change? What happens when occlusion occurs? What causes the cyclone to die out?

WAVE CYCLONES 120

Wave cyclone: vortex-like cyclone involving interaction of cold and warm air masses along sharply defined fronts.

Wave theory: concept of wave cyclone development based on interaction of cold and warm air masses along fronts. Developed by J. Berknes, about time of World War I.

Stages in life history of wave cyclone: polar front zone is trough of low pressure. Wave forms and deepens. Cold front and warm front emanate from low pressure center, take convex form in direction of motion. Cold and warm sectors develop. Precipitation intensifies along warm and cold fronts.

Warm air sector narrows; occlusion occurs. Cyclone dies as warm air source is cut off.

CYCLONES ON THE DAILY WEATHER MAP

How is a weather map constructed? How are pressure, fronts, winds, and precipitation areas shown?

Describe in detail the cyclone in an open stage. Include the following points: (1) shape of isobars; (2) isobars across the cold front; (3) wind patterns; (4) air mass movements; (5) wind shift along cold front; (6) precipitation areas; (7) following anticyclone; (8) cold and warm sectors.

Describe the cyclone in the occluded stage. What is the shape of the front? How has pressure changed in the storm center?

CYCLONES ON THE DAILY WEATHER MAP 121

Weather map constructed of isobars, frontal symbols, wind arrows, shows precipitation areas.

Cyclone in open stage: (1) isobars closed in oval shape; (2) isobars make sharp V crossing cold front; (3) winds form counter-clockwise in spiral; (4) warm, moist air flows into warm air sector, then over warm front; (5) sudden wind shift occurs as cold front passes; temperature drops abruptly; (6) precipitation zone broad over warm front, narrow along cold front; (7) anticyclone (high) follows cyclone, has clear skies; (8) northwest sector of cyclone is cold, southeast sector is warm.

Cyclone in occluded stage: low pressure intensifies. Occluded front recurves strongly toward northwest into low center. Precipitation falls in entire central region.

WAVE CYCLONE TRACKS

What direction do most wave cyclones take?

Where are most cyclones concentrated in the northern hemisphere?

What is a *cyclone family?* How do cyclones change in form from one end of the family group to the other?

WAVE CYCLONE TRACKS 124

Tracks run from west to east, follow common paths.

Cyclones concentrated in Aleutian and Icelandic lows.

Cyclone family: succession of wave cyclones tracking eastward along polar front while developing from open to occluded stage.

WAVE CYCLONES AND UPPER-AIR WAVES

Describe and explain the relationship between wave cyclones and the jet stream of an overlying upper-air wave.

How is an upper-air cut-off low related to a low-level cyclone?

WAVE CYCLONES AND UPPER-AIR WAVES 126

Relationship of wave cyclones to upper-air waves and jet stream: cyclones form beneath divergence zone of wave; anticyclones beneath convergence zone. Rising air in low-level cyclone is removed in upper-air divergence. Convergence in upper air causes sinking of air, feeding anticyclone.

Deep, intense occluded cyclone at low level is usually overlain by cut-off low of occluded Rossby wave.

TORNADOES

Describe a *tornado.* With what cloud type is it associated? What pressure conditions are present?

What is the *funnel cloud?* How wide is the funnel?

What wind speeds are typical of a tornado?

In what direction do tornadoes travel?

In what season do most tornadoes occur?

What is the major region of tornado occurrence in the U.S.?

What is a *waterspout?* Of what is it composed?

TORNADOES 127

Tornado: small, very intense wind vortex with extremely low pressure, formed beneath dense cumulonimbus cloud.

Funnel cloud: dark funnel-shaped cloud constituting the tornado vortex. Funnel 90–460 m (300–1500 ft) in diameter.

Wind speeds in tornadoes: up to 400 km (250 mi) per hr.

Direction of storm travel: west-to-east with moving cold front.

Season of occurrence: most occur in spring and summer months.

Region of U.S. occurrence: concentrated in Middle West, south. Rare west of Rockies.

Waterspout: intense perpendicular vortex of air and water droplets formed beneath a cumulonimbus cloud over tropical waters.

TROPICAL AND EQUATORIAL WEATHER DISTURBANCES

Describe general weather conditions in low latitudes. How strong is the Coriolis effect? Do air masses show strong contrasts? Are wave cyclones present? How important is convectional activity?

Describe an *easterly wave.* What pressure conditions are present? In what zone does it occur? In what direction does the wave travel? How fast? What weather is associated with an easterly wave?

TROPICAL AND EQUATORIAL WEATHER DISTURBANCES 129

Low latitude conditions: Coriolis effect weak; air masses have weak contrasts; sharply defined fronts missing; no intense wave cyclones. Convectional activity dominant in weak low pressure systems.

Easterly wave: slowly moving low-pressure trough in zone of tropical easterlies (trades), 5°–30° N and S. Wave travels westward 325–500 km (200–300 mi) per day, brings convectional showers and thunderstorms.

What is a *weak equatorial low?* What weather accompanies the low? In what zone is it found? How fast does it travel?

Weak equatorial low: weak, slowly moving low-pressure center accompanied by numerous convectional showers and thunderstorms. Forms close to ITC. Common in rainy season.

What is a *polar outbreak?* What air mass is involved? Where does the outbreak extend? What weather does it bring?

Polar outbreak: tongue of cold polar (cP) air, penetrating far into tropical latitudes. Brings rain squalls, unseasonal cold.

THE ASIATIC MONSOON

THE ASIATIC MONSOON 129

What are the features of the Asiatic monsoon circulation in January? Where is the ITC? What are the wind patterns over the East Indies? India and Burma? What is the effect on air mass movement?

In January, the ITC lies far to the south. Easterly trade winds at the surface and easterly upper-air winds are found over the East Indies. In India and Burma, north westerly surface winds and westerly upper-air winds are present. Southerly and southeasterly flow outflows of cP air masses produce a dry winter season.

What are the features of the Asiatic monsson circulation in July? Where is the ITC? What is the direction of upper air winds in the region? What is the direction of surface air flows? What is the effect of air mass movement?

In July, the ITC reaches northern India and lies across Burma, Thailand and the Philippines. Upper air easterlies dominate, from the ITC south toward the equator. Surface winds are from the west and southwest, opposite from winds aloft. mT air masses are steered westward from the Indian Ocean and China Sea into southern and eastern Asia.

TROPICAL CYCLONES

TROPICAL CYCLONES 131

What is a *tropical cyclone?* Describe the internal structure of the cyclone, including pressure, winds, and precipitation.

Tropical cyclone: intense cyclone of tropical zone, accompanied by high winds and heavy rainfall.

Define *hurricane.* In what region does it occur?

Hurricane: tropical cyclone of western North Atlantic and Caribbean region.

Define *typhoon.* In what regions does it occur?

Typhoon: tropical cyclone of western North Pacific Ocean and coastal waters of southeastern Asia.

What is the *central eye?* What weather conditions prevail in the eye?

Central eye: cloud-free central vortex of tropical cyclone.

Give the six regions of occurrence of tropical cyclones. What ocean has no tropical cyclones?

World distribution of tropical cyclones: (1) West Indies, Gulf of Mexico, Caribbean Sea; (2) western N. Pacific; (3) Arabian Sea, Bay of Bengal; (4) eastern N. Pacific off Mexico and Central America; (5) south Indian Ocean; (6) western S. Pacific (none in S. Atlantic).

Describe the characteristic track of tropical cyclones in the North Atlantic. Where do the storms originate? In what path and how fast do they travel? What finally happens to the cyclone?

Cyclone tracks in North Atlantic originate 10°–20° N; cyclones travel west, 10–20 km (6–12 mi) per hr, then turn northwest over coastal zone.

In what season do tropical cyclones occur? How is the ITC involved?

Season of occurrence: high-sun (summer) season of respective hemisphere, when ITC has moved farthest poleward.

What is the principal cause of destruction from tropical cyclones?

Cyclone destruction from high winds, waves, high water, river floods.

What is a *storm surge?*

Storm surge: rapid rise of coastal water accompanying arrival of cyclone.

FORECASTING TWO HURRICANES—GILBERT AND HUGO 133

Where did the National Hurricane Center (NHC) forecast Hurricane Gilbert would come ashore? How did this compare with a private weather forecaster? What was the outcome?

The NHC forecast that Hurricane Gilbert would come ashore at a point on the Mexican coast considerably south of Galveston, Texas. A private forecaster picked a location much nearer Galveston, and many residents of Galveston fled inland. The NHC forecast was correct.

Where did the NHC forecast landfall for Hurricane Hugo? Was the prediction correct?

The NHC forecast a landfall for Hurricane Hugo about 50 km (30 mi) southwest of Charleston, South Carolina. The prediction was correct.

What makes forecasting hurricane paths difficult?

Rapid changes in direction and strength in upper-air steering winds.

What was used to improved hurricane Hugo tracking?

The computer model called ETA was used.

REMOTE SENSING OF WEATHER PHENOMENA 134

What vehicles are used in remote sensing of weather phenomena?

Remote sensing of weather phenomena uses Nimbus satellites in polar orbits and geostationary orbits.

What kinds of weather data do remote sensing satellites obtain?

Satellites take TV photos and infrared images, record vertical temperature profile of atmosphere, measure ozone, water vapor, albedo, cloud cover, precipitation, and sea ice.

What is a *geostationary orbit?* What satellite uses such an orbit?

Geostationary orbit: satellite orbit that holds a fixed position over a selected point on the earth's equator. Example: Synchronous Meteorological Satellite (SMS).

What is an *ozone holes?*

Ozone hole: region of ozone depletion, often sensed by satellite remote sensing.

EL NIÑO AND THE SOUTHERN OSCILLATION 137

What is the meaning of the term *El Niño?*

El Niño: an invasion of warm surface waters off the Peruvian coast, replacing the usual cool upwelling Humboldt current; an event occurring every few years at the December solstice period.

Where does the warm water come from?

The warm surface water comes from the west as a surface current.

What is meant by the *Southern Oscillation?*

Southern Oscillation (SO): a reversal in the prevailing or normal barometric conditions in the equatorial and subtropical zone in a belt extending from the Indonesian/north Australian region to the eastern Pacific. (Combination of El Niño and SO gives the acronym ENSO.)

Describe the pressure reversal during SO.

During SO the usual low-pressure region centered on Darwin, northern Australia, develops high pressure, while the pressure in the region surrounding Tahiti deepens.

What effect has the SO?

The SO reversal causes the trades to weaken and disappear, replaced by an equatorial westerly wind system that causes warm surface water to be driven eastward against the Pacific shores of the Americas.

How was the prevailing jet stream pattern affected by ENSO?

ENSO was accompanied by a southerly shift in the track of the northern hemisphere jet stream.

List some of the widespread weather disturbances attributed to ENSO.

Effects of ENSO included intensive cyclonic storm activity in the western United States, torrential floods in the Andean highlands, drowning of Pacific coral reefs, drought in southern and southeastern Asia.

What term applied to such widespread global effects of ENSO?

The term "teleconnections" is applied to the widespread global effects of ENSO.

What is *La Niña*?

La Niña is the condition that occurs between El Niño events.

Give three characteristics of La Niña.

Three characteristics are: (1) sea surface temperatures fall to low levels; (2) Southeast tradewinds intensify; (3) warm surface water moves westward.

CHAPTER 7—SAMPLE OBJECTIVE TEST QUESTIONS

A. MATCHING

1. source region
2. warm, moist air
3. cold, dry air
4. warm, dry air
5. cold front
6. warm front
7. wave theory
8. hurricane
9. typhoon
10. tornado

a. ________ cP
b. ________ funnel clouds
c. ________ stratiform clouds
d. ________ Caribbean
e. ________ mE
f. ________ western Pacific
g. ________ cT
h. ________ air mass
i. ________ J. Berknes
j. ________ cumulonimbus

B. MULTIPLE CHOICE

1. The occluded front has

________a. warm air above cold air.

________b. cold air above warm air.

________c. cold air cut off from the ground.

________d. warm air at ground level in the middle.

2. In a wave cyclone, passage of the cold front brings

________a. a sudden rise in air temperature.

________b. a sudden shift in wind direction from west to east.

________c. a sudden shift in wind direction from east to west.

________d. is followed by falling barometric pressure.

3. Within the central eye of a tropical cyclone

________a. pressure is very high.

________b. winds are very strong.

________c. skies are heavily overcast.

________d. calm air prevails.

C. COMPLETION

1. Areas of land or ocean over which air masses develop definite properties are ________________ ________________________.

2. A front along which warm air is forced to rise along a gentle slope over cold air is a__________ ________________________front.

3. A succession of wave cyclones tracking eastward while developing from open to occluded stage is a ______________________________ .

4. A tongue of cold air, penetrating far into tropical latitudes is a ______________________________ ______________________________.

5. A rapid rise of coastal water accompanying arrival of a cyclone is a ______________________________ ______________________________.

CHAPTER 8

GLOBAL CLIMATE SYSTEMS

PAGE

CLIMATE AND CLIMATE CLASSIFICATION

What is the geographer's concept of climate? Why should the geographer be particularly interested in the ways in which climate influences plants and soils?

Define *climate.*

Define *climatology.*

CLIMATE AND CLIMATE CLASSIFICATION 141

Concept: geographer's view of climate as source of heat energy and water for needs of land plants. Climate acts upon plants and soil as an independent agent of control. Plants and soil, interacting in the life layer, respond to climate.

Climate: generalized statement of the prevailing weather conditions at a given place based upon statistics of a long period of record.

Climatology: the science of climate.

AIR TEMPERATURE AS A BASE FOR CLIMATE CLASSIFICATION

What role does air and soil temperature play in the life processes of organisms, particularly plants. How is survival involved in temperature extremes? What processes depend upon temperature?

AIR TEMPERATURE AS A BASE FOR CLIMATE CLASSIFICATION 141

Organisms respond to temperature of the medium that surrounds them. Plants cannot survive extremes of heat or cold. Rates of photosynthesis and respiration rise with increasing temperature, to optimum values.

THERMAL REGIMES

Explain the concept of *thermal regimes.* What is a regime? What controlling factors do the thermal regimes reflect? Name the six principal regimes.

Describe the equatorial regime. Where is it found? Are seasons developed? What is the mean monthly and annual temperature?

THERMAL REGIMES 141

Annual cycles of mean air temperatures fall into distinctive types, or *thermal regimes,* reflecting latitude and location with respect to continents and oceans.

Equatorial regime: uniformly warm, close to 27°C (80°F). No temperature seasons.

Describe the tropical continental regime. Where does it occur? What seasons are present? How large is the annual range?

Tropical continental regime: occurs in tropical deserts. Very hot at time of high sun; mild at low sun. Moderate annual range.

Describe the tropical west-coast regime. Why is the annual range so small? Why are temperatures so cool for this low latitude?

Tropical west-coast regime: weak annual cycle with temperatures exceptionally cool for low latitude. Cool ocean current offshore keeps temperatures down.

Describe the midlatitude west-coast regime. Why is the annual range small compared with interior locations?

Midlatitude west-coast regime: weak cycle persists into high latitudes on west coasts. Maritime effect; cool current off coast.

Describe the midlatitude and subarctic continental regimes. Why is the annual range so great? Why are winters so cold?

Midlatitude and subarctic continental regimes: large annual range, increases with higher latitude. Cold winters because of great net radiation deficit.

Describe the ice sheet regime. Compare it with the subarctic continental regime.

Ice sheet regime: extreme cold all year, but with strong annual cycle. Range not as great as subarctic continental regime.

Define *continentality.* Where will continentality be most strongly felt? Explain.

Continentality: tendency of large landmasses in midlatitudes and high latitudes to impose a large annual range on the temperature cycle.

Explain the marine influence upon temperature regimes. On which side of a continent will the marine influence be strongest on coastal climates?

Marine influence: opposite to continentality; is effect of large ocean bodies in suppressing the annual temperature cycle. Strongly felt on west coasts windward to westerlies.

PRECIPITATION AS A BASIS FOR CLIMATE CLASSIFICATION

PRECIPITATION AS A BASIS FOR CLIMATE CLASSIFICATION 142

Name seven precipitation regions. On what are the regions based? For each region, listed below, give amount of rainfall annually, latitude, prevalent air mass types, and type of precipitation:

Precipitation regions, based on annual total precipitation:

1. Wet equatorial belt

1. Wet equatorial belt; wet, over 200 cm (80 in.). mE air masses prevail. Convectional rainfall abundant.

2. Trade-wind coasts

2. Trade-wind coasts: 25°–30° latitude. Wet or very humid. 150–200 cm (60–80 in.) and over. mT air masses lifted in strong orographic effect.

3. Tropical deserts. What pressure belt do these deserts occupy? Why is the air dry?

3. Tropical deserts: less than 25 cm (10 in.). Belt of subtropical high pressure centered on tropics of cancer and capricorn. Subsiding cT air masses. Deserts extend off west coasts, far out over oceans.

4. Midlatitude deserts and steppes. Why are these areas dry?

4. Midlatitude deserts and steppes: under 10 cm (4 in.) in dry parts; steppes 25–50 cm (10–20 in.) cT air masses dominate. Moisture sources distant. Rainshadow effect important.

5. Moist subtropical regions

5. Moist subtropical regions: 25°–45° on southeast side of continent. 100–150 cm (40–60 in.) precipitation. mT air masses invade from subtropical high pressure cells on moist western sides.

6. Midlatitude west coasts. Why are these coasts so wet?

6. Midlatitude west coasts: 35°–65°. Precipitation over 200 cm (80 in.) where mountains lie near coast. Orographic effect on mP air masses in westerlies.

7. Arctic and polar deserts

7. Arctic and polar deserts: poleward of 60°. Cold cP and cA air masses hold little moisture. Mostly less than 30 cm (12 in.).

SEASONAL PATTERNS OF PRECIPITATION

What are the basic patterns of seasonal precipitation?

Name six seasonal precipitation types. For each type listed below, describe the annual cycle and the seasons: (See Figure 8.5, pg. 146.)

1. Wet equatorial type
2. Asiatic monsoon type
3. Wet-dry tropical type
4. Tropical desert type
5. Continental, summer-maximum type
6. West-coast, winter maximum type

SEASONAL PATTERNS OF PRECIPITATION 148

Precipitation patterns: (1) uniformly distributed; (2) summer (high-sun) maximum; (3) winter (low-sun) maximum.

Seasonal precipitation types:

1. Wet equatorial type: abundant rainfall in all months. May show one season with much greater rainfall. No dry period.
2. Asiatic monsoon type: strongly seasonal cycle, with peak rainfall in monsoon rainy season. Short dry season.
3. Wet-dry tropical type: very wet rainy season at high-sun alternates with long, very dry season at low-sun.
4. Tropical desert type: little precipitation in any month. May be weakly seasonal, with more rainfall at high-sun.
5. Continental, summer-maximum type: moderate to strong annual cycle with marked summer maximum.
6. West-coast, winter maximum type: may have dry summer; the winter maximum is always marked.

A CLIMATE SYSTEM BASED ON AIR MASSES AND FRONTAL ZONES

What is the concept of basing a climate system on air masses and their source regions? Does the system explain the climate, or merely describe it?

A CLIMATE SYSTEM BASED ON AIR MASSES AND FRONTAL ZONES 148

Concept: temperature and precipitation characteristics derive from air mass source regions and frontal zones. Explain climate characteristics in terms of properties of air masses and the nature of their interactions.

Name the three major climate groups. What is the basis of the grouping? For each group named below, give major air mass types in control, their source regions, and the general seasonal nature of the climates included:

Climate types, 13 in number, are grouped into three major groups as follows:

GROUP I. LOW-LATITUDE CLIMATES

Group I: Low-latitude Climates. Controlled by tropical and equatorial air masses. Includes source regions of the cT air masses in the subtropical high-pressure cells over continents, the zone of tropical easterlies (trades), and the ITC of the equatorial trough which is the source of the mE air masses. Climates of Group I lack a true winter season.

GROUP II. MIDLATITUDE CLIMATES

Group II: Midlatitude Climates. Largely under the influence of the polar front where interaction occurs between tropical (mT, cT) and polar (cP, mP) air masses. Wave cyclones are frequent at least at one season in the climates of Group II. Seasons are well developed and include both a summer and a winter.

GROUP III. HIGH-LATITUDE CLIMATES

Group III: High-latitude Climates.

Dominated by polar (cP, mP) and arctic (antarctic) (cA, cAA) air masses and including source regions of those air masses. *Arctic front zone* locally develops between polar and arctic air masses. Climates lack a summer season.

CLIMATE TYPES

CLIMATE TYPES 149

How many *climate types* are in each group?

Total of 13 *climate types,* or *climates;* Group I, four types; Group II, six; Group III , three.

What is a *climograph?* What data are used in the climograph?

Climograph: graph showing monthly mean values of both temperature and precipitation, for a given station, usually based upon many years of records.

DRY AND MOIST CLIMATES

DRY AND MOIST CLIMATES 150

Define *dry climate.*

Dry climate: climate in which annual evaporative losses of water from soil and plant cover exceed the annual precipitation by a wide margin.

What vegetation characterizes a dry climate?

Vegetation in dry climate is sparse, mostly shrubs or grasses.

Define *moist climate.*

Moist climate: climate with sufficient precipitation to maintain the soil in a moist condition through much of the year. Streams in a moist climate can flow year-around.

What vegetation characterizes a moist climate?

Vegetation in a moist climate consists of forest or tall-grass prairie.

What subtypes are recognized within the dry climates?

Subtypes of the dry climate:

Semiarid subtype (steppe subtype) (s): enough precipitation to support steppe grasslands.

Semidesert subtype (sd): transitional between semiarid and desert subtypes.

Desert subtype (d): extremely dry; small precipitation allows only growth of scattered plants.

What subtypes are recognized within the moist climates?

Subtypes of the moist climates:

Subhumid subtype (sh): annual evaporative losses of soil water nearly balance annual precipitation: transitional between dry and moist climates.

Humid subtype (h): enough precipitation to produce stream flow throughout the year and to support forest.

Perhumid subtype (p): heavy precipitation, copious stream flow.

KÖPPEN CLIMATE SYSTEM 155

How are the climates defined in the Köppen system?

Each climate is defined according to assigned values of temperature and precipitation. Each are computed on monthly and annual basis.

How are the Köppen system climates identified?

The climates are identified by shorthand code of letters with major groups, subgroups and further subdivisions.

CHAPTER 8—SAMPLE OBJECTIVE TEST QUESTIONS

A. MULTIPLE CHOICE

1. The tropical west-coast thermal regime

________a. has a large annual range.

________b. occurs in midlatitudes.

________c. has cold winters.

________d. has a small annual range.

2. The marine influence upon temperature regimes is most strongly felt

________a. on east coasts in midlatitudes.

________b. on west coasts in belt of westerlies.

________c. along the shores of the Arctic Ocean.

________d. behind mountain barriers in midlatitudes.

3. Annual precipitation on trade-wind coasts is usually

________a. 50–100cm (20–40 in.).

________b. 25–50 cm (10–20 in.).

________c. 100–150 cm (40–60 in.).

________d. 150 cm + (60 in.+).

4. Precipitation of the Asiatic monsoon type

________a. has a strong peak at time of high sun.

________b. has no dry season.

________c. has uniform distribution throughout the year.

________d. comes mostly from cT air masses.

5. Climates of Group I, Low-latitude Climates,

________a. have a winter season.

________b. lack a winter season.

________c. are wet throughout the year.

________d. are dry throughout the year.

B. COMPLETION

1. The tendency of large landmasses to impose a large annual range on the temperature cycle is called ____ .

2. The climate group in which there is strong interaction between polar and tropical air masses is known as ____ .

3. A graph showing monthly means of both temperature and precipitation is a ______________

______________________.

4. A semiarid climate subtype with sufficient precipitation to support short grasses is the______

______________________.

CHAPTER 9

THE SOIL-WATER BALANCE AND CLIMATE

PAGE

What concept stresses availability of water to plants?

Concept: availability of water to plants is a more important environmental factor than precipitation alone.

Define *hydrology*.

Hydrology: science of water as a planetary flow system in vapor, liquid, and solid states.

SURFACE AND SUBSURFACE WATER

SURFACE AND SUBSURFACE WATER 161

Define *surface water*. Name some forms of surface water.

Surface water: water of the lands flowing exposed (as streams) or ponded (as ponds, lakes, or marshes).

Define *subsurface water*.

Subsurface water: water held in soil and available to plants through their root systems; a form of subsurface water.

HYDROLOGIC CYCLE AND THE GLOBAL WATER BALANCE

HYDROLOGIC CYCLE AND THE GLOBAL WATER BALANCE 161

What is the *hydrologic cycle?* Describe the various pathways of the hydrologic cycle. In what states does the water exist?

Hydrologic cycle: interconnected system of pathways in which global free water moves and is stored in vapor, liquid, and solid states.

Define *runoff*. In what form does runoff travel? Where does it go?

Runoff: flow of water from continents to oceans by way of stream flow and ground water flow.

What is the *global water balance?*

Global water balance: a balance among the components of precipitation, evaporation, and runoff for the earth as a whole.

State the water balance equation.

Water balance equation: $P = E + G + R$, where P is precipitation, E is evaporation, G is net gain or loss of water in storage, and R is runoff, all in cu km/yr.

How is the equation simplified for application to long spans of time?

For long spans of time, storage gain or loss, G is constant and drops out: $P = E + R$.

HYDROLOGIC CYCLE AS A CLOSED MATERIAL FLOW SYSTEM

Describe a model of the hydrologic cycle as a closed material flow system

HYDROLOGIC CYCLE AS A CLOSED MATERIAL FLOW SYSTEM 162

Hydrologic cycle as a closed material flow system has three open subsystems, each with inputs and outputs forming a single closed system.

THE GLOBAL WATER BALANCE BY LATITUDE ZONES

Describe the water balance in the equatorial zone. What does this mean?

Describe the water balance in the subtropical belts. What does this mean?

Describe the water balance poleward of 40° latitude. What happens at high latitudes?

THE GLOBAL WATER BALANCE BY LATITUDE ZONES 162

The water balance in the equatorial zone shows an excess of precipitation over evaporation, yielding a net positive runoff. This means that water leaves the belt by streamflow, movement of ocean currents, or by movement of moist air masses.

In the subtropical belts, evaporation exceeds precipitation, and runoff is negative. This means that the belt receives water from other latitude belts by streamflow or ocean currents.

Poleward of 40° latitude, precipitation exceeds evaporation and runoff is positive. At high latitudes, values decline rapidly, reaching nearly zero at the poles.

GLOBAL WATER IN STORAGE

Where is most of the world's water held? Give percentages. How much is ground water?

How does the amount of surface water compare with the amount of ground water? How does the amount in stream flow compare with the amount in freshwater lakes?

GLOBAL WATER IN STORAGE 163

Most global water is in oceans (97.2%) and in glaciers (2.2%). Of remaining 0.6%, nearly all is ground water.

Surface water is very small amount (0.03%) and one third of that is salty. Stream flow holds very small fraction of that in freshwater lakes.

GLOBAL WATER BALANCE AND CLIMATE CHANGE

In what way might the global water balance be changed? Give a hypothetical example.

GLOBAL WATER BALANCE AND CLIMATE CHANGE 163

Atmospheric cooling on global scale could reduce precipitation and runoff, might increase water storage as snow and ice.

INFILTRATION AND RUNOFF

Define *infiltration*. Where does infiltration occur?

Under what conditions does overland flow occur?

INFILTRATION AND RUNOFF 163

Infiltration: absorption and downward movement of precipitation in soil layer.

Runoff by overland flow occurs when precipitation rate exceeds infiltration rate (covered in Chapter 16).

EVAPORATION AND TRANSPIRATION

When does evaporation of soil moisture occur? How thick a soil layer is affected?

Define *transpiration.* Where does transpiration take place?

Define the term *evapotranspiration.*

What is the *soil-water belt?*

What is the *intermediate belt?* Is water of the intermediate belt available to plants?

What is gravity percolation? Where does it take place? How far down does it extend?

EVAPORATION AND TRANSPIRATION 164

Evaporation of soil moisture occurs in dry weather, can affect upper 30 cm of soil in single dry season.

Transpiration: combined water loss through both soil evaporation and transpiration. Measured in same units as precipitation (cm/yr, in/yr).

Evapotranspiration: combined water loss from direct evaporation and transpiration of plants.

Soil-water belt: zone of soil-water available to plants.

Intermediate belt: zone below soil-water belt, too deep to supply capillary water to plants, i.e., too deep to be reached by plant roots.

Gravity percolation: downward movement of water under gravity force through soil-water and intermediate belts; may reach ground water zone below.

GROUND WATER AND STREAM FLOW

What happens to surplus water carried downward by gravity percolation?

Describe the paths of motion of ground water. Where does ground water energy?

When do streams receive water from overland flow?

GROUND WATER AND STREAM FLOW 164

Gravity percolation carries excess water down to ground water zone, eventually emerging in streams.

Ground water moves in deep paths, eventually emerges in streams, springs, lakes, or along ocean shoreline (see Chapter 16).

Streams are fed by overland flow in periods of heavy precipitation or rapid snowmelt.

WATER IN THE SOIL

Define *soil-water recharge.* What happens after the soil-water belt becomes saturated?

What happens in the soil-water zone after saturation has occurred? How is water held in the soil-water zone?

Define *storage capacity (field capacity)* What units are used? Upon what factor does storage capacity depend?

WATER IN THE SOIL 164

Soil-water recharge: restoring of depleted soil-water by infiltration of precipitation. When soil-water belt is saturated, excess water percolates downward to intermediate belt.

Drainage of soil-water follows saturation. Gravity removes all except capillary water capable of being held in soil pores.

Storage capacity (field capacity): maximum capacity of soil to hold water against pull of gravity. Measured in units of water depth (cm, in.). Storage capacity depends upon soil texture.

THE SOIL-WATER CYCLE

Describe the typical annual cycle of soil-water increase and decrease for a humid climate in midlatitudes. Give conditions and changes for each season as follows:

Early spring. Is storage capacity reached?

Summer. What causes depletion of soil-water?

Autumn. Why does evapotranspiration decrease?

Winter. Why is evapotranspiration nearly zero?

Define *water surplus.* What happens to surplus water?

Define *water deficit.* When does a deficit usually occur?

THE SOIL-WATER CYCLE 165

Soil-water increases and decreases in annual cycle through seasons. Example is for humid climate in midlatitudes where cold winter alternates with warm summer.

Early spring: soil-water close to storage capacity.

Summer: high air temperatures and heavy evapotranspiration depletes soil-water storage.

Autumn: evapotranspiration declines as plants become dormant, air temperatures decline.

Winter: soil-water frozen, evapotranspiration near zero.

Water surplus: water disposed of by runoff or by percolation to ground water zone because storage capacity of soil is full.

Water deficit: difference between soil-water present and storage capacity of soil, a condition typical of a dry period or dry season.

THE SOIL-WATER BALANCE

What is the *soil-water balance*?

State the soil-water balance equation.

Define *soil-water storage* (S).

Define *actual evapotranspiration* (water use).

Define *potential evapotranspiration* (water need). What conditions of plant cover are assumed to exist? What conditions of water availability are assumed? Under what circumstances is water need a real, rather than a hypothetical value?

Define *soil-water shortage.* In terms of crop growth, what is the significance of the soil-water shortage?

THE SOIL-WATER BALANCE 165

Soil-water balance: balance among the component terms of the soil-water budget; namely precipitation (P), evapotranspiration (E), change in soil-water storage (G), and water surplus (R).

Soil-water balance equation: $P = E + G + R$ Units are cm/month or cm/yr.

Soil-water storage (S): actual quantity of water held in the soil-water belt at any given instant; stated in cm, water depth equivalent.

Actual evapotranspiration (water use) (Ea): actual rate of evapotranspiration; represents the real situation.

Potential evapotranspiration (water need) (Ep): an ideal, or hypothetical, rate of evapotranspiration estimated to occur from a complete canopy of growing plants if they were continuously supplied with all the soil water they could use. Not hypothetical where precipitation is sufficiently great or irrigation water is supplied in sufficient amounts.

Soil-water shortage (D): difference between water use and water need. Represents the quantity of water that must be furnished by irrigation to sustain greatest possible rate of crop growth.

SOIL-WATER BALANCE AS AN OPEN MATERIAL FLOW SYSTEM

Describe a model of the soil-water balance as an open material flow system

Define *surface detention.*

SOIL-WATER BALANCE AS AN OPEN MATERIAL FLOW SYSTEM 166

Soil-water balance can be modeled as an open material flow system, tracing the input of water from precipitation, several storages within the system, transformations of state by evaporation, and multiple system outputs to atmosphere, to surface water, and to ground water.

Surface detention: temporary storage of precipitation in surface depressions of soil surface, when precipitation rate exceeds infiltration rate.

A SIMPLE SOIL-WATER BUDGET

What is a *soil-water budget?* What items or terms are included in the soil-water budget?

Define *storage withdrawal* (–G).

Define *storage recharge* (+G).

Describe an idealized sample soil-water budget typical of a humid midlatitude climate. Give conditions in each season as follows:

Winter. Is there a water surplus? Explain.

Spring. Why does water need increase? Is soil water taken from storage? What value of storage capacity is assumed to exist?

Summer. How do plants respond to reduction in soil water availability? Does a shortage develop? Why?

Autumn. How does water need change in this season? Explain. When is soil-water recharge begun?

What scientist developed the principles of the soil-water budget? How were they applied?

A SIMPLE SOIL-WATER BUDGET 166

Soil-water budget: an accounting system evaluating amounts of precipitation, evapotranspiration, water storage, water deficit, and water surplus at a given time and place.

Storage withdrawal (–G): depletion of stored soil water during periods when potential evapotranspiration (water need, Ep), exceeds precipitation (P); it is difference between water use (Ea) and precipitation.

Storage recharge (+G): restoration of stored water during periods when precipitation (P) exceeds potential evapotranspiration (water need, Ep).

Described below is an idealized, or sample soil-water budget typical of humid midlatitude climate. Field capacity is assumed to be 30 cm (12 in.).

Winter: water surplus exists. Water need and water use are both small because of cold soil temperatures, dormancy of plants.

Spring: water need begins to exceed precipitation. Plants begin to draw upon soil water in storage.

Summer: storage withdrawal increases. Plants conserve soil water by reducing their evapotranspiration. Water use falls well below water need; shortage develops.

Autumn: water need declines rapidly and soon drops below precipitation. Soil-water storage is first recharged. When soil storage is full, surplus sets in.

C. Warren Thornwaite developed principles of soil-water budget; applied them to need for irrigation of crops during growing season.

CALCULATING A SIMPLE SOIL-WATER BUDGET

How is a soil-water budget calculated? What data are used?

CALCULATING A SIMPLE SOIL-WATER BUDGET 168

Calculation of soil-water budget uses monthly means, which may be averaged over a long period of record at a given station. Simple addition and subtraction give changes in G, monthly values and annual totals of R and D.

GLOBAL RANGE OF WATER NEED

Why is there a great global range in water need? Does the total annual need vary from place to place? Is there an annual cycle of water need?

Explain how latitude acts as a general control of water need. In what latitude zones is water need greatest? Why? Where is water need least? Explain. What has air temperature to do with global differences in water need?

What controls seasonal variations in water need: Is the seasonal range greater in lower latitudes than in higher latitudes?

What is distinctive about the annual water-need cycle in subarctic and arctic zones?

GLOBAL RANGE OF WATER NEED 169

Water need (potential evapotranspiration, Ep) shows a wide global range in terms of (1) total annual quantity, and (2) annual cycle of water need according to season of year.

Latitude as a control of water need: annual total water need is greatest in equatorial and tropical zones; declines to minimum in arctic zone. Temperature is the major control acting through latitude.

Seasonal variations are controlled by seasonal temperature cycles. Seasonal range increases with higher latitude.

In subarctic and arctic zones, several consecutive winter months have zero water need, when soil water is frozen.

SOIL-WATER BALANCE AS A BASIS OF CLIMATE CLASSIFICATION

What advantages has a climate classification using the soil-water balance?

SOIL-WATER BALANCE AS A BASIS OF CLIMATE CLASSIFICATION 171

Advantages of a climate system using the soil-water balance: (1) it is a quantitative system using available station data of global scope; (2) it gives information of direct value in assessing conditions favorable and unfavorable to growth of natural plant cover and agricultural crops; (3) it is the system used by soil scientists to establish a modern soil classification.

SOURCE OF DATE

What source of data is available for a water-balance climate system?

SOURCE OF DATA 171

Data source consists of soil-water balances of over 13,000 stations of global extent, computed by Thornthwaite Laboratory of Climatology; same data used by U.S. Dept. of Agriculture for soil classification.

WATER NEED AS AN INDICATOR OF AVAILABLE HEAT

How can available heat (sensible heat) be expressed through the soil-water balance?

Explain the use of total annual Ep in distinguishing the three major climate groups.

How are Low-Latitude Climates, Group I, defined?

How are Midlatitude Climates Group II, defined?

How are High-Latitude Climates, Group III, defined?

WATER NEED AS AN INDICATOR OF AVAILABLE HEAT 171

Total annual potential evapotranspiration (annual water need, Ep) is directly related to heat available to plants throughout the year or in a growing season.

Major climate groups (see Chapter 8) can be defined by total annual value of Ep as follows:

Group I, Low-Latitude Climates: climates with total annual Ep greater than 130 cm.

Group II, Midlatitude Climates: climates with total annual Ep ranging from 130 cm down to 52.5 cm.

Group III, High-Latitude Climates: climates with total annual Ep less than 52.5 cm.

DEFINING DRY AND MOIST CLIMATES

Define a *dry climate.*

Define a *moist climate.*

In what way is the dry-moist climate boundary related to natural vegetation?

DEFINING DRY AND MOIST CLIMATES 171

Dry climate: climate in which total annual soil-water shortage, D, is 15 cm or larger, while water surplus, R, is zero.

Moist climate: climate in which the soil-water shortage, D, is less than 15 cm.

In midlatitude zone, dry-moist climate boundary coincides approximately with boundary between steppe grasslands (short-grass prairie) and tall-grass prairie.

SUBDIVISIONS OF THE DRY CLIMATES

Give name, symbol, and definition of each of the three subtypes of the dry climates.

SUBDIVISIONS OF THE DRY CLIMATES 172

Symbol	*Subtype*	*Definition*
s	Semiarid	At least two months in which soil-water storage, S, is equal to or exceeds 6 cm.
sd	Semidesert	Fewer than two months with S greater than 6 cm, but at least one month with S greater than 2 cm.
d	Desert	No month with S greater than 2 cm.

SUBDIVISIONS OF THE MOIST CLIMATES

Give name, symbol, and definition of each of the three subtypes of the moist climates.

SUBDIVISIONS OF THE MOIST CLIMATES 172

Symbol	*Subtype*	*Definition*
sh	Subhumid	Soil-water shortage, D, is greater than zero but less than 15 cm, when there is no water surplus, R. Otherwise, D is greater than R, when R is not zero.
h	Humid	R is 1 mm or greater but less than 60 cm. (R is always greater than D.)
p	Perhumid	R is 60 cm or greater.

CLIMATES WITH STRONG MOISTURE SEASONS

How are climates with strong moisture seasons distinguished?

CLIMATES WITH STRONG MOISTURE SEASONS 173

Two climates have strong moisture seasons, i.e., alternating very dry and very moist seasons. These climates show both a substantial soil-water shortage (D) and a substantial soil-water surplus (R) (see Climates (2) and (3)).

CHAPTER 9—SAMPLE OBJECTIVE TEST QUESTIONS

A. MATCHING

1.	surface water	a. _______	saturation
2.	transpiration	b. _______	gravity
3.	ground water	c. _______	evaporation
4.	surplus	d. _______	lake
5.	percolation	e. _______	runoff

B. MULTIPLE CHOICE

1. The water balance equation for the earth as a whole includes

________a. infiltration, evaporation, and runoff.

________b. percolation, transpiration, and runoff.

________c. percolation, evaporation, and precipitation.

________d. precipitation, evaporation, and runoff.

2. The amount of global total water held in the oceans and glaciers combined is about

________a. 90%.

________b. 99%.

________c. 69%.

________d. 89%.

3. Storage capacity represents the quantity of water held

________a. in the soil against the force of gravity.

________b. in the ground water zone.

________c. in the intermediate zone.

________d. in the saturated soil.

C. COMPLETION

1. Water of the lands held in the soil or rock below the surface is ______________________________ water.

2. The combined water loss through both soil evaporation and transpiration of plants is ______ _________________________.

3. The zone below the soil-water belt, too deep to supply water to plants is the ________________ _________________________ belt.

4. Difference between water use and precipitation is the ____.

5. Difference between soil water present and storage capacity of the soil is the________________ _________________________.

6. Difference between water use and water need is the____.

CHAPTER 10

LOW-LATITUDE CLIMATES

	PAGE

THE LOW-LATITUDE CLIMATES

THE LOW-LATITUDE CLIMATES 174

In what latitude do the low-latitude climates lie?

Low-latitude climates lie in equatorial zone (10°N to 10°S) and tropical zone (10° to 15° N and S).

What wind and pressure belts do these climates occupy?

Equatorial belt of doldrums and ITC, tropical easterlies (trades), and portions of oceanic subtropical highs.

WET EQUATORIAL CLIMATE (1)

WET EQUATORIAL CLIMATE (1) 174

Rainfall: How is rainfall distributed through the year? What is the annual total? Is there a seasonal difference?

Rainfall: heavy in all months. Annual total over 200 cm (80 in.), and often over 250 cm (100 in.) May have one season of heavier rainfall.

Location: Give latitude range.

Location: astride equator, lat 10°N to 10°S.

What are major regions of occurrence?

Major regions of occurrence: South America (Amazonia), Africa (Congo basin), East Indies.

Temperatures: Compare monthly and annual means. How large is the annual range?

Temperatures: both monthly and annual means are close to 27°C (80°F). Very small annual range.

Describe the soil-water budget.

Water need is uniformly high through the year; precipitation is large in nearly all months; nearly all months show a substantial water surplus. The total annual water surplus is very large.

Daily Range in Air Temperature

Daily Range in Air Temperature 175

Describe the daily range in air temperature .

Daily range in temperature is from 8 to 11 C°, which is much larger than annual temperature range of only about 1 C°, or even less.

MONSOON AND TRADE-WIND LITTORAL CLIMATE (2)

Rainfall: When does heavy rainfall occur? When is the low-rainfall period? Give annual total rainfall.

Location: In what latitude range is the climate found?

What are the major regions of occurrence?

Temperatures: What cycle occurs? How large is the annual range? How does the annual mean compare with the wet equatorial climate?

Describe the soil-water budget of the monsoon and trade-wind littoral climate.

Asiatic Monsoon Variety

Rainfall: Describe the features of this climate.

Temperatures: What cycle occurs? How large is the annual range?

Describe the soil-water budget.

The Rainforest Environment

What are the outstanding qualities of the *low-latitude rainforest environment?*

Describe the flow of streams in the rainforest environment.

Describe the soil conditions if the rainforest environment.

Describe the natural (native) vegetation of the low-latitude rainforest.

MONSOON AND TRADE-WIND LITTORAL CLIMATE (2) 175

Rainfall: heavy rainfall in season just following high-sun solstice (June-September in northern hemisphere). Period of low rainfall after low-sun solstice (March-April). Annual total large, usually over 200 cm (80 in.).

Location: 5° to 25° N and S along narrow east coastal strips windward to trades.

Major regions of occurrence: E. sides of Central and S. America, Caribbean Islands, Madagascar, Indochina, Philippines, W. India coast, Burma coast.

Temperatures: marked annual cycle with maximum in high-sun period. Annual range about twice that of wet equatorial climate. Mean about same as wet equatorial climate.

Water need shows a well developed annual cycle with maximum in high-sun period. Precipitation has strong annual cycle with peak September-November. Result is a very large water surplus in the rainy season and a modest shortage in the short dry season.

Asiatic Monsoon Variety 176

Rainfall: shows an extreme peak of rainfall in high-sun period; dry season is two to three months long.

Air temperatures show only a very weak annual cycle, cooling a bit during the rains. Annual range is small at this low latitude.

The soil-water budget shows a small shortage in the dry season and a large surplus during rainy months.

The Rainforest Environment 177

Low-latitude rainforest environment: combines extreme uniformity of monthly mean air temperature with large annual total rainfall, great annual water surplus, and high soil-water storage throughout the year.

Stream flow is copious most of the year. Rivers are important arteries of travel. Towns and cities are situated on river banks.

Bedrock is deeply weathered; soil usually thick, red in color and rich in oxides (Oxisols). Soil is poor in plant nutrients, not fertile for grasses and grain crops.

Natural vegetation is *broadleaf evergreen rainforest,* consisting of great forest trees with broad leaves that are not shed seasonally.

What is important about the relationship of this forest to plant nutrients?

Rainforest trees quickly recycle nutrients from surface layer of the soil, where they have been released by decay of fallen leaves and branches.

Describe the flora and fauna of the low-latitude rainforest.

Low-latitude rainforests have great number of tree species in only a small area. Fauna (animal assemblage) is also very rich. Most animals live in forest canopy; these include birds, bats, insects, reptiles, and invertebrates. Examples: climbing monkeys.

Plant Products and Food Resources of the Rainforest

Plant Products and Food Resources of the Rainforest 178

Name some rainforest products in the categories of lumber, drugs, and food.

Rainforest products: lumber: mahogany, ebony, balsawood. Drugs: quinine, cocaine. Foods: cocoa (cacao plant).

What is the forest source of natural rubber?

Natural rubber, made from sap of rubber tree, originally from South America. Modern production largely from SE Asia.

Name some of the starchy staples found in the rainforest.

Starchy staples: manioc (cassava), yam, taro (poi), banana, plantain.

What tree of the rainforest environment provides both food and fiber?

Coconut palm tree provides food (meat of nut, dried as copra), vegetable oil, fiber from leaves (roof thatch), lumber from trunk.

WET-DRY TROPICAL CLIMATE (3)

WET-DRY TROPICAL CLIMATE (3) 179

Rainfall: When does the wet season occur? What is the total annual rainfall? Which months are practically rainless? What air masses bring the rains? What air masses produce the dry season?

Rainfall: strongly developed wet season (monsoon is Asia) in high-sun season following solstice. Total: 100–150 cm (40–60 in.). Long dry season at time of low sun. Three or four rainless months typical. Rains produced from mE air mass when ITC is over the region; dry season by cT air mass.

Location: What latitude range is occupied by this climate?

Location: 5° to 25° N and S (10° to 30° in Southeast Asia).

What are the major regions of occurrence?

Major regions of occurrence: India, Indochina, W. and S. Africa, S. America, N. Australia.

Temperatures: describe the annual temperature cycle. When is the hot season? When is the cool season? Give mean temperatures of maximum month and minimum month.

Temperatures: annual cycle well developed with hot season just before onset of rains, cool season at low-sun.

Maximum month: 27°–32°C (80°–90°F).

Minimum month: 16°–21°C (60°–70°F).

Describe the soil-water budget of the wet-dry tropical climate.

Water need shows strong annual cycle, peaking in hot season (April-June) just prior to monsoon rains (June-September), which bring a water surplus. Large water shortage develops in early months of year.

The Savanna Environment — 181

Describe the characteristic native vegetation of the wet-dry tropical climate.

Plants adapted to survive a long drought season, then to grow rapidly in a short season; the general term for this is *rain-green vegetation.*

What two basic types of vegetation comprise the rain-green savanna vegetation?

Two types. *Savanna woodland,* an open forest with widely spaced trees and coarse grasses in open spaces between. Thorny trees and shrubs in patches, with grasses between.

What name is given by geographers to the environment described above?

Savanna environment describes this combination of climate and vegetation.

What conditions of stream flow characterize this environment?

Streams (rivers) are completely or almost dry in dry season, flow full to banks in rainy season.

Can the rains be counted on to fall copiously each year?

Rains may fail in a given year, or series of years, bringing drought and crop failures.

What kind of soil conditions are present in the savanna environment?

Red soils of low fertility are present widely on uplands, but large areas of fertile soil exist where windblown dust has accumulated. Fertile, silty alluvial soils can be found on river floodplains.

Animal Life of the African Savanna — 182

Describe the animal life of the African savanna.

Carnivores, such as lion, leopard, and hyena, feed on herbivores, such as wildebeeste and zebra. Other herbivores—rhino and elephant—find defense in great size and strength or thick hide.

What effect has the dry season on this assemblage of animals?

In dry season, grasses and surface water dry up, leaving sparsely distributed water holes. Competition for water and food is severe. Some herbivores migrate long distances in search of food and water.

What is the major threat to this animal population?

Animal population is threatened by loss of native vegetation habitats through expansion of farming and by killing of game animals for hides, horns, and tusks.

Agricultural Resources of the Savanna Environment — 182

Describe the cultivation of rice in Southeast Asia.

Rice is dominant staple cereal of Southeast Asia. It is grown in climates (2), (3), (6). Fields (paddies) must be flooded in rainy season and become dry for harvest.

Name other important food crops of the savanna environment.

Other food crops: sugarcane, sorghum (kaffir corn, guinea corn), peanut.

Food and Woodland Resources of the African Savanna — 182

Describe *bush-fallow farming.* Where is it practiced? What crops are cultivated?

Bush-fallow farming: practiced in moister zones of savanna woodland with long wet season. Trees are cut and burned, ash left on ground. Cultivation follows until soil fertility is lost, then tree growth resumed and cycle repeated. Crops are grains, yams, soybeans, sugarcane.

Describe the agricultural system of the dry savanna grassland and thorntree savanna.

In drier areas of savanna climate, short wet season allows cultivation of sorghum, millet, peanuts, and corn.

What other agricultural activity takes place in this same farming region?

In same region, there exists a shifting cattle culture, practiced by nomadic groups that follow rain-green forage.

Name some other plant products of the African savanna.

Other plant products: trees used as firewood and for construction; cashew-nuts, kapok.

DRY TROPICAL CLIMATE (4)

DRY TROPICAL CLIMATE (4) 183

Rainfall: How much annual rainfall has the desert subtype? the steppe subtype? How great is the potential evaporation for the year?

Rainfall: desert subtype (4d), less than 4 cm (10 in.) annual total, little in any month. Steppe subtype (4s), 25–76 cm (10–30 in.) annual total. Potential for evaporation very large; up to 200 cm (80 in.) per year. Steppe rainfall occurs in short rainy season at high sun.

Location: What latitude range is spanned by the dry tropical climate? How are the steppe zones situated with respect to the desert zones? What air mass source region coincides with the desert?

Location 15°–25° N and S, centered on tropics of cancer an capricorn. Steppe subtype forms transitional zone on equatorward side of desert. Desert subtype occupies source regions of cT air masses. Semidesert subtype (4sd) is transition zone between steppe and desert.

What are the major regions of occurrence?

Major regions of occurrence: Sahara—Arabia—Thar desert belt of N. Africa and S. Asia, Australia, S. America, S. Africa.

Temperatures: Describe the annual temperature cycle. Give mean of the hottest month; of the coldest month. How large is the annual range?

Temperatures: very high temperatures at time of high sun:

Mean of hottest month: 32°–38°C (90°–100°F).

Coldest month mean: 10°–16°C (50°–60°F).

Large annual range: 17–22 C° ª30–40 F°).

Describe the soil-water budget of the dry tropical climate.

Water need show strong annual cycle following temperature curve. Very large total annual water need. Precipitation is much less than water need and water use in every month. Result is water shortage in every month, near-zero soil-water storage in every month, and a very large total annual soil-water shortage.

Western Littoral Subtype (4dw)

Western Littoral Subtype (4dw) 184

Rainfall:

Rainfall: often less than 10 cm (4 in.) yearly total.

Location: Give latitude range for western littoral subtype climate.

Location: found on narrow west-coast strips, latitude 15°–30° N and S.

Major region of occurrence:

Major region of occurrence: Atacam Desert, Namib Desert.

Temperatures: How large is the annual range? How do temperatures compare with those at inland areas at the same latitude? What is the range of the monthly mean temperatures? Explain this temperature regime. Are fogs common? Why?

Temperatures: range very small for tropical latitude. Exceptionally cool for tropical latitude. Monthly means stay mostly between 16° and 21° C (60° and 70° F). Cool, upwelling currents offshore keep air cool. Fog frequent.

Semiarid Subtype (Tropical Steppe) (4s)

Rainfall:

Location:

Major areas of occurrence:

Temperatures:

Describe the soil-water budget for this climate.

Semiarid Subtype (Tropical Steppe) (4s) 185

A prolonged hot, dry period of four or more consecutive months precedes the short wet season, which starts at about the time of high-sun solstice.

Lies as a transition between wet-dry tropical (3) and dry tropical semidesert (4sd) climates.

Principally located in Africa, Asia, Australia, and South America.

Warm, building to very hot at high-sun before the onset of the rainy season.

The soil-water budget shows a small surplus in the wet season, but it is not sufficient to recharge soil water. A large shortage prevails for the rest of the year.

The Tropical Desert Environment

Describe desert rainfall and its effects.

What becomes of stream flow in desert basins with no outlet?

What is a desert pavement?

Describe typical desert plants.

What kinds of native vegetation are found in the tropical desert environment?

The Tropical Desert Environment 186

Desert rainfall is unreliable, sporadic and often brief and torrential, causing flash flooding and deposition of bouldery debris.

Streams arriving at a low place with no outlet deposit clay and silt, while the evaporating water leaves a salt deposit. Salt flats result Shallow salt lakes may be permanent features.

Barren desert land surfaces may accumulate a loose cover of close-fitting pebbles to form a desert pavement, easily disturbed by wheels of vehicles.

Many kinds of desert plants are hardleaved or spiny, resisting water loss. Many annuals flower and seed only every few years, when rainfall occurs.

Tropical desert vegetation is of two general classes: *dry desert*, with widely scattered perennial plants; *semidesert* with abundant woody shrubs or grass clumps; *thorntree semidesert* (Africa) with scattered thorny trees and shrubs.

RAINFALL VARIABILITY OF THE LOW-LATITUDE CLIMATES

Define *rainfall variablity:*

What are three classes of rainfall variability that apply to low-latitude climates? Give an example climate of each class.

RAINFALL VARIABILITY OF THE LOW-LATITUDE CLIMATES 186

Rainfall variability: the degree of reliability of rainfall from year to year. It is important because it measures the hazards forces in raising food crops within the climate zone.

Three classes: (1) low variability (wet equatorial climate (1)); (2) moderate variability (monsoon and trade-wind littoral (2)) wet-dry tropical climate (3); high variability (dry tropical climate (4))

HIGHLAND CLIMATES OF LOW LATITUDES

What are the effects of increased altitude? How is insolation affected by altitude increase? How is heating and cooling affected?

Daily air temperature: How does daily range change with increasing altitude?

Monthly mean air temperature: How does it change with altitude?

Precipitation: How is it affected by increased altitude?

HIGHLAND CLIMATES OF LOW LATITUDES 188

Effects of increased altitude: decrease in air density allows more intense insolation. Faster heating and cooling of ground surface and lower air layer, than at low altitudes.

Daily air temperatures: daily range increases with increased altitude.

Monthly mean air temperatures: decline with increased altitude.

Precipitation: usually increases with increased altitude, due to orographic effect.

MOUNTAIN AGRICULTURE IN THE HIGH EQUATORIAL ANDES

Describe mountain agriculture in the high equatorial Andes.

MOUNTAIN AGRICULTURE IN THE HIGH EQUATORIAL ANDES 188

In the high equatorial Andes of Peru and Bolivia or Altiplano (3200–4300 m), corn, wheat, barley, and potatoes are grown.

CHAPTER 10—SAMPLE OBJECTIVE TEST QUESTIONS

A. MULTIPLE CHOICE

1. The dry tropical climate (4) owes its aridity to

   ________a. location in the ITC throughout the year.

   ________b. continually rising air in high-pressure cells.

   ________c. constant, strong trade-winds that keep moisture out.

   ________d. subsiding cT air mass over the tropical zone.

2. Of the low-latitude climates listed below, which has the greatest annual water surplus?

   ________a. wet-dry tropical (3).

   ________b. wet equatorial (1).

   ________c. dry tropical (4).

   ________d. monsoon (2).

3. Which of the following is not a starchy staple food plant of the low-latitude rainforest environment?

   ________a. manoic.

   ________b. yam.

   ________c. taro.

   ________d. quinine.

4. Broadleaved evergreen forest of the rainforest obtain their mineral nutrients largely from

   ________a. a fall of dust particles.

   ________b. a thin surface layer of the soil where leaves and branches quickly decay.

   ________c. a layer 1 to 2 meters below the soil surface.

   ________d. dissolved minerals in the rain that falls.

B. COMPLETION

1. Vegetation of the savanna environment is described as ____________________ vegetation.

2. Kind of agriculture practiced in the savanna environment of Africa: ____________________ ____________________.

3. Name the natural native vegetation of the low-latitude rainforest ____________________ ____________________.

4. Class of native vegetation occurring in Africain the semidesert climate subtype (4sd) of the dry tropical climate: ____________________ .

CHAPTER 11

MIDLATITUDE AND HIGH-LATITUDE CLIMATES

THE MIDLATITUDE CLIMATES

	PAGE
THE MIDLATITUDE CLIMATES	**190**

What latitude zones are occupied by the midlatitude climates?

Midlatitude climates occupy the midlatitude zone and a large proportion of the subtropical zone; they extend into the subarctic zone in Europe.

What zones of air masses and fronts do the midlatitude zones occupy?

Midlatitude climates lie generally in the polar-front zone where mT air masses interact with mP and cP air masses.

What kinds of weather systems are typical of the midlatitude climates?

In the zone of westerly winds, traveling cyclones and anticyclones move through these climate zones from west to east.

DRY SUBTROPICAL CLIMATE (5)

DRY SUBTROPICAL CLIMATE (5) **190**

Temperatures: How do temperatures differ from that of the dry tropical climate?

Temperature range greater than for dry tropical climate (4). Has distinct cool or cold season at time of low sun. Midlatitude wave cyclones may enter area in low-sun (winter) season.

Location: What latitude range is occupied by this climate?

Latitude range: 25°–35° N and S. Adjoins and grades into the dry tropical climate (4) on equatorward side.

What climate subtypes are recognized?

Same three subtypes as in dry tropical climate: 5s, 5sd, 5d.

Describe the soil-water budget of the dry subtropical climate.

Similar in many respects to dry tropical climate, but with smaller total annual water need and smaller soil-water shortage. Rains in a moist cool season bring a small amount of soil-water recharge at time of low sun.

The Subtropical Desert Environment

What North American desert exemplifies the subtropical desert environment? Describe the plants and animals of this desert.

The Subtropical Desert Environment 191

The Mojave Desert of southeastern California is hot in summer but has a distinct winter season. Plants include the saguaro cactus, Joshua tree, ocotillo, and creosote bush. The numerous desert mammals are largely nocturnal in habit; many inhabit burrows by day.

MOIST SUBTROPICAL CLIMATE (6)

Precipitation: How is precipitation distributed throughout the year in the humid subtropical climate? What is the annual total? What kind of rainfall is dominant in summer? What is special about the summer rainfall of Southeast Asia?

Location: On which side of a continent is this climate found? between what latitudes? Between what two climates is this climate transitional? What air masses influence this climate?

Temperatures: Describe the general features of the moist subtropical climate?

What are the major regions of occurrence?

Describe the soil-water budget of the moist subtropical climate.

East-Asian Monsoon Variety

What is distinctive about the east-Asian variety of the moist subtropical climate (6)? What happens in winter?

The Moist Subtropical Forest Environment

Describe the flow of streams in this environment.

What is the native forest type of much of this environmental region?

MOIST SUBTROPICAL CLIMATE (6) 192

Precipitation: ample in all months, but typically with a summer maximum. Annual total: 125–150 cm (50–60 in.), or higher. Convectional showers and thunderstorms common in summer. Southeast Asia shows strong summer rainfall peak (monsoon).

Location: eastern sides of continents, 20°–35° N and S. Transitional between low-latitude and midlatitude climates. Position on moist west sides of subtropical high-pressure cells, subject to mT air mass invasions.

Temperatures: summers warm; Winters mild with persistent high humidity and no winter month mean temperature below freezing.

Major regions of occurrence: S.E. United States, Formosa, S. Japan, Uruguay, parts of Brazil and Argentina, E. coast of Australia.

Water need shows strong annual cycle with peak in summer. Precipitation is substantial in all months. A very small soil-water shortage develops in summer. Annual water surplus is large.

East Asian Monsoon Variety 193

The east-Asian monsoon variety of the moist subtropical climate has a strong summer precipitation maximum produced by the summer monsoon. In winter, precipitation is lower when a dry air mass from interior Asia dominates.

The Moist Subtropical Forest Environment 193

Stream and rivers flow copiously through much of the year. Flooding can be severe from tropical cyclones in summer and fall months.

Broadleaf evergreen forest was the native forest of much of this area before deforestation occurred. Evergreen oak and magnolia are typical trees. Pine forest occupies sandy soils of the region.

Describe the properties of the soils of this environment.

Soils are mostly heavily leached and poor in nutrients needed for grains and cereals. Soils are typically yellow or red in color. Severe soil erosion has been a major problem on cultivated lands.

Agricultural Resources of the Moist Subtropical Forest Environment

Agricultural Resources of the Moist Subtropical Forest Environment 194

What is the principal cereal crop produced in this environment in Eastern Asia, esp. southern China and Japan? Describe the cultivation system.

Rice is principal cereal crop in southern China, Japan, and Taiwan. Elaborate terrace systems are used to create rice fields. High yields are obtained by use of fertilizers.

What important plant product of Southeast Asia is not grown in the United States?

Tea, widely cultivated in Japan and China, is not grown in the United States.

What are the principal field crops of this environment in the southern United States?

In southern U.S., principal field crops are corn, cotton, peanuts, and tobacco.

Name other important agricultural products of the southern U.S.

In southern U.S., cattle production is becoming important in areas of sandy soils, along with tree farming (pines).

MEDITERRANEAN CLIMATE (7)

MEDITERRANEAN CLIMATE (7) 195

Precipitation: What precipitation conditions prevail in summer? Is rainfall ample in winter?

Precipitation: summers extremely dry, often rainless for several months. Rainfall (rarely snow) is ample in winter.

Location: On what side of a continent is this climate found? Between what latitudes?

Location: west sides of continents, 30°–45° N and S.

What are the major regions of occurrence?

Major regions of occurrence: central and S. California, coastal zones surrounding Mediterranean Sea. W and S. Australia, Chile coast, Cape Town region of S. Africa.

Temperatures: Describe in general terms the temperature found in the narrow coastal zone of the Atlantic and Pacific; then for the Mediterranean lands and inland locations.

Temperatures: in the narrow coastal zones on Pacific and Atlantic oceans: summers cool; winters mild. On the Mediterranean lands and inland locations: summers warm. Annual temperature range is moderate.

Describe the soil-water budget of the Mediterranean environment.

Water need shows strong annual cycle with peak in summer. Precipitation shows strong annual cycle with peak in winter. Result is large soil-water shortage throughout summer. Recharge occurs in winter, but may not be enough to produce a water surplus.

The Mediterranean Climate Environment

The Mediterranean Climate Environment 197

In what ways is the Mediterranean climate environment attractive to human occupation?

Mediterranean climate environment attracts humans through its benign climate, with mild sunny winters.

What major drawback to human occupation is seen in this environment?

Large annual water shortage in the soil budget limits human occupation. Water supplies must be imported from distant sources.

Describe the soils of this environment as to their fertility.

Soils of valley and lowland areas are highly fertile, and with irrigation provide high yields of crops.

Define *sclerophylls*.

Sclerophylls: hardleaved evergreen trees and shrubs capable of enduring a summer drought.

What special environmental hazards beset humans in southern California?

Wildfires sweeping through dry brush *chaparral* are major hazard. When rains follow, soil is swept into stream channels to flood low areas, leaving thick layers of mud and boulders.

What special form of environmental degradation affected the Mediterranean lands of Europe, North Africa, and the Near East.

In the Mediterranean lands of Europe, severe soil erosion over several centuries time caused by deforestation and over-grazing, has depleted soils from hillsides.

What special agricultural products are produced in lands bordering the Mediterranean Sea?

Besides cereals, the Mediterranean lands produce citrus fruits, grapes, and olives. Cork of the cork oak is a unique industrial product.

What kinds of agricultural products are important in southern California?

In southern California, irrigated alluvial soils yield vegetable crops, sugar beets (for sugar), and forage crops (alfalfa).

What environmental problem is associated with prolonged irrigation of lowland soils?

Lowland soils of California in several areas are degraded by waterlogging and salt accumulation.

MARINE WEST-COAST CLIMATE (8)

MARINE WEST-COAST CLIMATE (8) 199

Precipitation: Describe the general features of the annual cycle of precipitation.

Precipitation: precipitation ample in all months, but with marked decrease in summer months. Rainfall is much higher in orographic situations. Heavy winter snows in higher latitudes and altitudes.

Location: Between what latitude limits is this climate found? What belt is this in terms of weather patterns and air masses?

Location: west coasts, 35°–60° N and S. Belt of midlatitude wave cyclones. Moist mP air masses yield much precipitation.

What are the major regions of occurrence?

Major regions of occurrence: W. coast of N. America, W. Europe and British Isles, Victoria, Tasmania, New Zealand, Chile.

Temperatures: Describe the temperature ranges in general terms.

Temperatures: summers and winters cool depending upon latitude; Annual range small.

Describe the soil-water budget of the marine west-coast climate.

Water need shows moderately strong annual cycle with summer maximum, but with small amounts throughout mild, rainy winter. Very small soil-water shortage in summer; large water surplus in winter.

The Marine West-Coast Environment

The Marine West-Coast Environment 200

Describe agricultural soils of the marine west-coast environment.

Agricultural soils of the marine west-coast environment in Europe, cultivated for centuries, retain moderate fertility but require fertilizers and lime.

Why are soils over much of this climate region unsuited to agriculture?

Widespread mountainous terrain, heavily scoured by ice sheets and glaciers of the Ice Age over large parts of the marine west-coast climate region, bears poor-quality soils, unfit for crop agriculture.

Describe the native vegetation of this environmental region.

Needleleaf forests cover the mountainous areas. Broadleaf, "summergreen" decidious forests were native to lowland areas in areas in Europe, but are largely replaced by cultivation.

Describe the agricultural development of western Europe in this climate region.

In western Europe, lowlands of the marine west-coast climate region have been intensively used for centuries for crop farming, orchards and forests.

What are the principal plant products of mountainous areas in this climate region in North America?

Dense forests of needleleaf trees (redwood, fir, cedar, hemlock, spruce) in mountainous areas of the Pacific Northwest are a major economic resource.

Describe the water resource of this climatic region.

Large water surpluses of the mountainous areas of the marine west-coast climate are a major resource for agricultural and industrial use and can be exported to distant regions of water need.

DRY MIDLATITUDE CLIMATE (9)

DRY MIDLATITUDE CLIMATE (9) 201

Precipitation: Describe the general features of the annual precipitation cycle of the dry midlatitude climate.

Precipitation: climate is dry with tendency to summer maximum caused by sporadic invasions of maritime air masses.

Location: Where are these dry climates located? Between what latitudes? What causes the deserts? In what belt of winds are these dry climates located? Of what air masses are these areas the source regions?

Location: continental interiors, 35°–55° N. Strongly developed on rainshadow (east) side of Cordilleran and Andean ranges in zone of westerly winds. Source regions of dry cT (summer) and cP (winter) air masses.

What are the major regions of occurrence?

Major regions of occurrence: W. North America, Eurasian interior.

Temperatures: Generally describe the temperatures of the dry midlatitude climate.

Temperatures: summers warm; winters cold depending on latitude. Great annual range.

Describe the soil-water budget of the dry midlatitude climate.

Water need shows large annual cycle with sharp peak in summer and several consecutive months of zero water need in cold winter. Precipitation also shows summer maximum, but a substantial shortage develops in summer. Recharge through winter not sufficient to produce a surplus.

The Dry Midlatitude Environment

The Dry Midlatitude Environment 202

How does dryness of climate relate to soil fertility?

Low precipitation and a continentality of temperature result in highly fertile soils; they retain nutrient elements called "bases," used by grasses and grain crops. These soils are alkaline in acid balance.

What is the native vegetation of this environment?

Native vegetation of the semiarid subtype of this climate is *short-grass prairie,* it consists of perennial grasses that can endure severe seasonal drought.

What is the name for this vegetation type in Asia?

In central Asia, the expanses of short-grass prairie are called *steppes;* the environment as a whole is called *steppe.*

What name is given to the class of soils typical of the short-grass prairie? Describe this soil.

Soils of the short-grass prairie are called *Mollisols.* These soils of loose texture, easily tilled, and brown to pale brown in color. A lower layer of this soil has excess calcium carbonate in the form of rocklike nodes.

Describe the role of wheat as an agricultural product of the short-grass prairie.

Wheat is the most important grain crop of the short-grass prairie region in North America and in the Ukraine region of Russia; also important in northern China.

MOIST CONTINENTAL CLIMATE (10)

MOIST CONTINENTAL CLIMATE (10) 203

Precipitation: Generally describe the precipitation pattern of the moist continental climate.

Precipitation: summer maximum marked, especially in eastern Asia. Decreases inland toward subhumid subtype.

Location: In what hemisphere is this climate found? Between what latitudes? What frontal zone is found in this climate region? What air masses are dominant in summer? in winter?

Location: northern hemisphere only, 30°–55° N, (Europe, 45°–60° N). Occupies polar-front zone between source regions of cP and mT or cT air masses.

What are the major regions of occurrence?

Major regions of occurrence: central and E. North America, N. China, Korea, N. Japan, central and E. Europe.

Temperatures: Describe the temperatures found in the moist continental climate.

Temperatures: summers warm depending on latitude; winters cold; annual range very great.

Describe the soil-water budget of the moist continental climate.

Water need shows strong annual cycle with a summer peak and near-zero values in winter. Precipitation is ample in all months. A very small shortage develops in summer. Annual water surplus is large.

The Moist Continental Forest and Prairie Environment

The Moist Continental Forest and Prairie Environment 205

What is the response of vegetation and soil in the humid part of the moist continental environment?

In humid part of the moist continental climate forest is the native vegetation, while soils show leaching out of the nutrient base and are acid in chemical balance.

What types of forest are found in this region?

In the northern portion, where acid sandy soils occur, is evergreen needleleaf forest; this grades southward into deciduous forest.

What effects of the Ice Age are seen here?

Great ice sheets of the Pleistocene Ice Age strongly influenced landforms and soil by eroding and transporting bedrock and soil.

Describe the climate and vegetation change encountered as one travels westward in interior North America.

In the Middle West, starting in Illinois and continuing west into Nebraska, climate becomes less humid (subhumid subtype, 10sh). Here soils grade into fertile Mollisols (prairie soils), formerly covered by *tall-grass prairie.*

What kind of agriculture is practiced in the more southerly parts of the moist continental climate?

Where land surfaces are relatively flat, crop farming for cereals (corn, wheat, rye, oats, barley), sugar beets and soybeans is intense.

What is the corn belt?

Corn belt: belt of intensive cultivation of corn and other crops, used as animal feed, in what was formerly the tall-grass prairie of the U.S. Midwest.

HIGH-LATITUDE CLIMATES

HIGH LATITUDE CLIMATES 207

In what areas are the high-latitude climates located?

High-latitude climates are almost entirely situated in the northern hemisphere lands of North America and Eurasia.

What air masses and frontal zones characterize the high-latitude climates?

Maritime polar (mP) interact with continental polar (cP) and arctic (A) air masses in the arctic front zone.

What characteristics of the soil-water budget apply to these climates?

Total evapotranspiration is low (less than 50 cm) and monthly values fall to zero for several consecutive months in winter when soil water is frozen.

BOREAL FOREST CLIMATE (11)

BOREAL FOREST CLIMATE (11) 207

Precipitation: Generally describe the annual precipitation cycle of the boreal forest climate.

Precipitation: maximum in summer. Little in the winter months. Annual total ranges from small, in interiors, to large, near oceans.

Location: In what latitude range is the climate found? In what hemisphere? What air mass source region is found here?

Location: 50° to 70° N (none in southern hemisphere). Continental interiors; source regions of cT air masses.

What are the major regions of occurrence?

Major regions of occurrence: central and W. Alaska, Canada, S. Greenland, Iceland, Eurasia.

Temperatures: Generally, what are the temperatures of the boreal forest climate?

Temperatures: winters severely cold; summers mild; largest annual range of all climates.

Describe the soil-water budget of the boreal forest climate.

Water need shows sharp summer peak, but several consecutive winter months have zero value and annual total is small. Precipitation also shows a summer peak, and soil-water shortage is very small. Melting snows in spring bring a large water surplus.

The Boreal Forest Environment

The Boreal Forest Environment 208

In what ways did the glaciation of the Ice Age affect the boreal forest environment?

Glaciation was intense over most of the region of the boreal forest environment, strongly eroding bed rock and transporting rock debris long distances. Many lakes and bogs were formed. Soils are poorly developed and mostly infertile.

What is peat? How is it used?

Peat is partly decomposed plant matter formed in bogs. It has been used as a low-grade fuel in northern Europe.

What is the natural vegetation of the boreal forest climate region?

Needleleaf forest is the native vegetation of the boreal forest climate; largely pine, spruce, and fir, but with deciduous larch forest typical of Siberia.

What is the *taiga?*

Taiga: cold woodland near the northern limit of tree growth; it consists of widely spaced low trees (black spruce) and intervening areas covered by mosses and lichens.

What plant products of economic importance are found in the boreal forest region?

Needleleaf forests yield valuable pulpwood logs, floated down rivers to lumber mills.

TUNDRA CLIMATE (12)

TUNDRA CLIMATE (12) 209

Precipitation: What is the total annual precipitation in the tundra climate? What variations are found?

Precipitation: mostly less than 30 cm (12 in.), but may be up to 50 cm (20 in.) in moister easterly locations, and much higher locally in orographic position.

Location: Where is the tundra climate found? At what latitude? In what frontal zone is this climate located? What air mass source regions lie on either side?

Location: lands fringing Arctic Ocean, mostly poleward of 60°–75° N and S. Extends south to 55 ° N latitude in eastern Canada. Located in arctic front zones, between source regions of cP or mP and cA air masses.

Temperatures: Describe the temperatures of the tundra climate in general terms.

Temperatures: no summer; winters severely cold; annual range large, but less than continental subarctic climate.

Describe the soil-water budget of the tundra climate.

Water need shows sharp summer peak and a winter period of eight consecutive months of zero value. Total annual water need is very small. Precipitation occurs in all months and only a very small shortage develops in summer. A substantial water surplus is developed in winter months; held frozen as snow until spring thaw.

The Arctic Tundra Environment

The Arctic Tundra Environment 210

What does the term *tundra* mean?

Tundra: a special class of natural vegetation and a particular environmental region, typical of a treeless landscape in a subarctic climate.

Describe the soils and vegetation of the arctic tundra.

Arctic tundra soils are stony and contain much humus; in summer they are water-saturated. Trees exist only as small shrublike forms. Other plants are grasses, sedges, and lichens.

Describe the animal life of the arctic tundra.

Animal life of the arctic tundra consists of few species. Mammals include caribou, musk oxen, wolverine, fox, polar bear, rabbits, lemmings. Insects that bite mammals are present in vast numbers in the summer. A lichen, "reindeer moss," is grazed by several mammal species. Waterfowl are abundant in summer.

ICE-SHEET CLIMATE (13)

ICE-SHEET CLIMATE (13) 212

Precipitation: Describe the precipitation expected in the ice-sheet climate.

Precipitation: very small, almost all snow.

Location: Where is the ice-sheet climate located? What are the three main areas?

Location: Greenland Ice Sheet, arctic sea ice, Antarctic Ice Sheet. Latitude range 65°–90° N and S. Source regions of cA and cAA air masses.

Temperatures: Compare ice-sheet temperatures with those of other cold climates. What is the mean of the coldest month? Where is it coldest?

Temperatures: lowest annual average of any climate; no month has mean above freezing. Much colder in high interior ice-sheet locations than near coast and on sea ice.

Describe the soil-water budget of the ice-sheet climate.

Water need is effectively zero in all months because monthly mean temperatures are well below freezing. Precipitation is scanty and is almost entirely in form of snow, which accumulates as glacial ice.

The Ice-Sheet Environment

The Ice-Sheet Environment 212

Describe the ice-sheet environment.

Ice-sheet environment of Antarctica is extremely hostile to humans and most animal and plant species, hence almost totally devoid of life except at the margins where habitat is near.

CHAPTER 11—SAMPLE OBJECTIVE TEST QUESTIONS

A MULTIPLE CHOICE

1. The moist subtropical climate (6) is located in the latitude range

________a. 15°–20°. ________c. 20°–35°.

________b. 45°–60°. ________d. 5°–15°.

2. The Mediterranean climate (7) is found on

________a. the west sides of continents.

________b. the east sides of continents.

________c. windward coasts in the trade-wind belt.

________d. poleward of 50° latitude.

3. The boreal forest climate (11) is characterized by

________a. uniform cold throughout the year.

________b. frequent invasions of mP air masses in winter.

________c. an extremely great annual range of temperature.

________d. the lowest air temperatures on earth in winter.

4. A zone of lichen-woodland along the northern fringe of the boreal forest is called the

________a. tundra. ________c. permafrost.

________b. taiga. ________d. frigid zone.

B. COMPLETION

1. Midlatitude climate with sufficient moisture to support grasslands, but not forests: ____________________________________.

2. Major class of soils characterized by brown color and loose texture, with calcium carbonate nodules in lower part: ________________________________.

3. Black organic substance formed by partial decomposition of forest foliage: ____________________________________.

CHAPTER 12

MATERIALS OF THE EARTH'S CRUST

PAGE

What is the lithosphere?

The lithosphere is a general term for the solid, mineral earth realm of the planet.

What are *landforms?*

Landforms: configurations of the land surface taking distinctive forms and produced by natural processes.

Define *geology*.

Geology: the science of the solid earth.

COMPOSITION OF THE EARTH'S CRUST

COMPOSITION OF THE EARTH'S CRUST **214**

Define the earth's *crust*.

Earth's *crust:* is the thin outermost layer of the earth.

Name in order of abundance the eight most abundant elements in terms of percent by weight. Which are metals? (See Figure 12.1, pg. 214.)

Eight most abundant elements, percent by weight: oxygen 47%, silicon 28%; together make up 75% of crust. Metals: aluminum 8.1%, iron 5.0%, calcium 3.6%, sodium 2.8%, potassium 2.6%, magnesium 2.1%

ROCKS AND MINERALS

ROCKS AND MINERALS **215**

Define *mineral.* Give the three important attributes of a mineral.

Mineral: a naturally occurring inorganic substance, usually having a definite chemical composition and a characteristic atomic structure.

Define *rock.*

Rock: an aggregate of minerals in the solid state. Most rocks consist of two or more minerals.

Name the three major rock classes.

Major rock classes: *igneous, sedimentary*, and *metamorphic*.

Define *igneous rock*

Igneous rock: rock formed of mineral matter solidified from a high-temperature molten state, i.e., from a *magma.*

What is *magma?*

Magma: moblie, high-temperature molten state of rock.

Define *sedimentary rock.*

Sedimentary rock: rock formed of layered accumulations of mineral particles derived from preexisting rocks or produced by organic activity.

Define *metamorphic rock.*

Metamorphic rock: igneous or sedimentary rock changed physically and chemically through heat or pressure, or through introduction of mineral substances by moving solutions.

THE SILICATE MINERALS

THE SILICATE MINERALS 216

Define *silicate minerals.* Besides silicate, what elements are usually present?

Silicate minerals: minerals containing silicon (Si) and oxygen (O) atoms linked together in the crystal structure. Most silicate minerals also contain one, two, or more of the following: aluminum (Al), iron (Fe), calcium (Ca), sodium (Na), potassium (K), magnesium (Mg).

Give composition and density of the following silicate minerals: (See Figure 12.2, pg. 215.)

Principal silicate minerals abundant in igneous rocks:

Quartz

Quartz: silicon dioxide (SiO_2), density 2.6 gm/cc.

What are the *aluminosilicates?*

Aluminosilicates: silicate minerals containing aluminum as an essential element. (The next five mineral groups are aluminosilicates.)

Feldspars

Feldspars: general term for an aluminosilicate group consisting of silicate of aluminum and one or more of the metals K, Na, Ca.

Potash feldspar

Potash feldspar: aluminosilicate with potassium as the dominant metal, $(K,Na)Si_3O_8$. Density 2.6 gm/cc

Plagioclase feldspar

Plagioclase feldspar: aluminosilicate group starting with sodic plagioclase, $NaAlSi_3O_8$, and grading in a continuous series to calcic plagioclase, $CaAl_2Si_2O_8$. Plagioclase of intermediate composition has both Na and Ca. Density 2.6–2.8 gm/cc.

Mica group

Mica group: complex aluminosilicates characterized by perfect cleavage in one set of parallel planes. One variety, biotite, has K, Mg, Fe, and water in its composition. Density 2.9 gm/cc.

Amphibole group

Amphibole group: complex aluminosilicates of Ca, Mg, Fe. Example: hornblende, a black mineral, density 3.2 gm/cc.

Pyroxene group

Pyroxene group: complex aluminosilicates of Ca, Mg, Fe. Example: augite, black color, density 3.3 gm/cc.

Define *density.*

Density: mass of substance (matter) per unit of volume; stated in gm/cc. (Density of pure water is 1 gm/cc.)

How do densities of the above-listed silicate minerals differ from one another?

Densities of the above-listed silicate minerals range in increasing order from 2.6 (quartz) to 3.3 (olivine).

Which of the silicate minerals are *felsic minerals?*

Felsic minerals: quartz and the feldspars, silicate minerals of light color and comparatively low density (2.6–2.8 gm/cc).

Which are *mafic minerals?*

Mafic minerals: group of minerals, mostly silicates, rich in Mg and Fe, dark in color, and of comparatively great density (2.9 to 3.3 in the case of the silicates).

What two common minerals, not silicates are included with the mafic minerals?

Magnetite (Fe_3O_4) and ilmenite ($FeTiO_3$) are also mafic minerals; their densities are $4\frac{1}{2}$–$5\frac{1}{2}$ gm/cc.

SILICATE MAGMAS

SILICATE MAGMAS 216

What is *silicate magma?*

Silicate magma: magma of silicate composition with included volatile elements and compounds. Temperature 500–1200°C.

Describe the cooling of a silicate magma.

Cooling of magma leads to crystallization of silicate minerals in complex series of interactions.

VOLATILES IN MAGMAS

VOLATILES IN MAGMAS 217

What are *volatiles?*

Volatiles: elements and compounds normally existing in the gaseous or liquid state under atmospheric conditions of temperature and pressure.

Where do volatiles come from?

Volatiles are dissolved in magma, reach earth's surface by *outgassing* after magma has crystallized.

Define *outgassing.*

Outgassing: process of exudation of water and other gases from earth's crust through volcanoes, to become part of the earth's atmosphere and hydrosphere.

Name the important volatiles in the earth's atmosphere and hydrosphere. (See Table 12.1, pg. 217.)

Important volatiles in atmosphere and hydrosphere: water (H_2O), carbon as CO_2 gas, sulfur S_2, nitrogen, (N_2), argon (A), chlorine (Cl_2), hydrogen (H_2), fluorine (F_2)

TEXTURES OF IGNEOUS ROCKS

TEXTURES OF IGNEOUS ROCKS 217

What is meant by *rock texture?*

Rock texture: crystal size and arrangement of mixed sizes in igneous rocks.

What effect does rate of cooling of magma have on texture of igneous rock?

Slow cooling of magma gives coarse texture; fast cooling gives fine texture or a glass.

What is *intrusive igneous rock?* What texture does it have?

Intrusive igneous rock: rock formed of magma solidified within the crust, enclosed by older solid rock. It has coarse texture where the rock mass is large and has cooled slowly.

What is *extrusive igneous rock*? What texture does it have?

Extrusive igneous rock: rock formed of magma that has reached the earth's surface, coming into direct contact with air or water. It has fine-grained or glassy texture.

What is *lava?*

Lava liquid magma that pours out upon the earth's solid surface, often taking the form of stream-like masses.

What are *obsidian, scoria,* and *pumice?*

Natural forms of volcanic glass are *obsidian,* a dense black glass, *scoria,* a spongelike lava with large gas cavities, and *pumice,* a fine-textured variety of glassy scoria.

What is *volcanic ash?* What general name applies to such material?

Volcanic ash: finely divided extrusive igneous rock blown out under gas pressure from a volcanic vent; it is a form of *tephra.*

Define *tephra.*

Tephra: collective term for all size grades of solid igneous particles blown from a volcanic vent by gas pressure.

CLASSIFICATION OF IGNEOUS ROCKS

CLASSIFICATION OF IGNEOUS ROCKS 218

Give the composition of five major intrusive igneous rocks and their important extrusive equivalents.

1. *Granite*

   Rhyolite

2. *Diorite*

   Andesite

3. *Gabbro*

   Basalt

4. *Peridotite*

5. *Dunite*

Major types of intrusive igneous rocks, arranged in order from felsic through mafic to untramafic types, with extrusive equivalents:

1. *Granite:* felsic igneous rock composed largely of quartz, potash feldspar, and plagioclase feldspar, with minor amounts of biotite and hornblende.

   Rhyolite: extrusive equivalent of granite.

2. *Diorite:* felsic igneous rock composed dominantly of plagioclase feldspar and pyroxene, with some amphibole and biotite mica.

   Andesite: extrusive equivalent of diorite.

3. *Gabbro:* mafic igneous rock composed largely of pyroxene and plagioclase feldspar, with variable amounts of olivine.

   Basalt: extrusive equivalent of gabbro.

4. *Peridotite*: ultramafic igneous rock composed largely of pyroxene and olivine.

5. *Dunite:* ultramafic igneous rock composed largely of olivine.

Define *felsic rock.*

Felsic rock: igneous rock composed largely of felsic silicate minerals.

Define *mafic rock.*

Mafic rock: igneous rock composed largely of mafic silicate minerals.

Define *ultramafic rock.*

Ultramafic rock: igneous rock composed largely of pyroxene and olivine.

What range of densities is covered by the felsic, mafic, and ultramafic igneous rocks?

Densities of igneous rocks: felsic, 2.7–2.8; mafic, 3.0; ultramafic, 3.3.

FORMS OF IGNEOUS ROCK BODIES

FORMS OF IGNEOUS ROCK BODIES 218

What are *plutons?*

Plutons: intrusive igneous rock bodies.

What is *batholith?*

Batholith: the largest of the intrusive igneous rock bodies.

Define *sill.*

Sill: pluton in the form of a plate where magma forced its way between two horizontal rock layers.

Define *dike:*

Dike: a near-vertical thin layer of pluton formed by the spreading apart of a vertical rock fracture.

What are *volcanoes?*

Volcanoes: conical dome-shaped structures built by the emission of lava from a restricted vent in the earth's surface.

What are *lava flows?*

Lava flows: igneous material that reaches the earth's surface and pours out in a tongue-like manner from the volcano.

DISINTEGRATION AND DECAY OF IGNEOUS ROCK

DISINTEGRATION AND DECAY OF IGNEOUS ROCK 219

In what way is the surface environment unfavorable to the preservation of igneous rock?

Surface environment—air, water—not favorable to preservation of silicate minerals and igneous rocks. Presence of water, CO_2, and free oxygen at low temperatures causes chemical change in minerals.

What is *bedrock?*

Bedrock: solid, unaltered rock.

What is *regolith?*

Regolith: layer of mineral particles overlying the bedrock and derived by weathering of the underlying bedrock or transported from another location by a fluid medium (fluid agent). Regolith is a source of sediment.

SIZE GRADES OF MINERAL PARTICLES

SIZE GRADES OF MINERAL PARTICLES 219

What is *sediment?*

Sediment: finely divided mineral matter and organic matter derived directly or indirectly from preexisting rock and from life processes.

What is *abrasion?* What processes are involved?

Abrasion: grinding action of streams, waves, glacial ice, or wind on exposed bedrock surfaces. Particles thus freed by be crushed between larger grains.

Name the major size grades, giving diameter limits in mm.

Grade sizes and limits used by U.S.D.A.:

Grade name:	Diameter limits
Sand	2.0 –0.05 mm
Silt	0.05–0.002 mm
Clay	below 0.002 mm

What are *colloids*? What is a colloidal suspension?

Colloids: mineral particles finer than about 0.01 micron, capable or remaining in suspension indefinitely in water when dispersed.

What shapes have mineral particles of sediment?

Larger particles (sand and larger) tend to be rounded by abrasion; mineral silt and clay particles often highly angular or platelike.

How does surface area of mineral grains vary with particle size?

Decreasing particle size brings enormous increase in surface area of particles in a given volume of sediment. Important in favoring chemical reactions.

CHEMICAL ALTERATION OF SILICATE MINERALS

CHEMICAL ALTERATION OF SILICATE MINERALS 220

What is *mineral alteration?*

Mineral alteration: chemical reactions that change the composition of silicate minerals of igneous rock (primary minerals) to new compounds (secondary minerals), stable in the surface environment.

Define *primary minerals.*

Primary minerals: original silicate minerals of igneous rock.

Define *secondary minerals.*

Secondary minerals: mineral alteration products produced by chemical weathering.

What is *oxidation?* With what elements in minerals does oxygen combine?

Oxidation: union of free oxygen with metallic elements in minerals. Oxygen combines easily with calcium, magnesium, and iron.

Explain *carbonic acid reaction.*

Carbonic acid reaction: action of weak solution of carbonic acid upon susceptible minerals.

What is *hydrolysis?*

Hydrolysis: chemical union of water molecules with minerals to from new, more stable compounds.

MINERAL PRODUCTS OF HYDROLYSIS AND OXIDATION

MINERAL PRODUCTS OF HYDROLYSIS AND OXIDATION 220

What is a *clay mineral?* What is the origin of clay minerals? What are the properties of clay minerals?

Clay mineral: class of secondary minerals, produced by alteration of silicate minerals, having plastic properties when moist. Colloidal flakes of clay mineral hold layers of water molecules.

What is *kaolinite?* What is its color? For what is it used?

Kaolinite: clay mineral produced by alteration of potash feldspar. White when pure; used in ceramics. Composition: $Al_2Si_2O_5(OH)_4$.

What is *Bauxite*?

Bauxite: mixture of several clay minerals, consisting largely of *sesquioxide of aluminum* and water.

Define *sesquioxide of aluminum.*

Sesquioxide of aluminum: combination of 2 atoms of Al with 3 atoms of O: Al_2O_3.

How is *illite* formed?

Illite: clay mineral formed by alteration of feldspars and mica (muscovite); a hydrous aluminosilicate of K.

How is *montmorillonite formed?*

Montmorillonite: clay mineral formed by alteration of feldspar, mafic minerals, or volcanic ash; a hydrous aluminosilicate of Fe, Mg, Na. Swells greatly when it absorbs water.

What is *vermiculite?*

Vermiculite: clay mineral similar in composition and structure to montmorillonite, but does not swell when wetted. Formed by hydrolysis of mafic aluminosilicates.

Describe *hematite* and *limonite.*

Two forms of iron sesquioxide are *hematite* (Fe_2O_3) and *limonite* ($2Fe_2O_3 \cdot H_2O$), formed by oxidation of iron in mafic minerals. Hematite gives brown and reddish colors to soil and rocks.

How is free *silica* formed in soil and regolith?

Hydrolysis of silicate minerals releases *silica* (SiO_2), a secondary mineral of same composition as quartz.

What form does silica take in soil and rock?

Redeposited silica forms *chalcedony,* made up of minute silica crystals.

CLASTIC SEDIMENT

CLASTIC SEDIMENT 221

What are *clastic sediments?*

Clastic sediments: sediments derived directly as particles broken from parent rock source.

What are *nonclastic sediments?*

Nonclastic sediments: sediments formed of newly created mineral matter precipitated from inorganic chemical solutions or by organic activity.

What are *pyroclastic sediments?*

Pyroclastic sediments: accumulations of volcanic materials (tephra) transported by wind or water.

What are *detrital sediments?*

Detrital sediments: fragments derived by the decomposition and/or disintegration of preexisting rocks of any type.

What are the most abundant mineral constituents in clastic sediments?

Most abundant mineral constituents in clastic sediments: quartz, particles of unaltered parent rock, also feldspar and micas. Clay minerals make up bulk of fine-grained sediments.

What is *sorting?* How does it occur?

Separation of clastic sediments into grade size groups or layers is *sorting;* caused by different rates of transport and fallout in moving water or wind.

What causes colloidal clays to settle out as sediment layers?

Flocculation, the clotting together of clay colloid particles, results from encounter with salts of seawater. Clotted particles settle to bottom.

CLASTIC SEDIMENTARY ROCKS

CLASTIC SEDIMENTARY ROCKS 223

What is *lithification?*

Lithification: compaction, hardening, and cementation of sediment to form sedimentary rock.

What is *volcanic breccia?*

Volcanic breccia: a crude mixture of large and small pyroclastic fragments.

What is *tuff?*

Tuff: volcanic ash transported by wind or water and deposited in layers.

What is *sedimentary breccia?*

Sedimentary breccia: coarse, sedimentary rock.

What is *conglomerate?*

Conglomerate: common rock consisting of pebbles, cobbles or boulders in a fine-grained matrix of sand or silt.

Describe *sandstone.*

Sandstone: sedimentary rock formed of sand grades of sediment, cemented by SiO_2 or calcium carbonate ($CaCO_3$).

What kinds of mineral particles make up sandstone?

Sandstone is commonly made up of grains of quartz, and sometimes feldspar, mica, or heavy minerals; it may also consist of sand-sized rock particles containing several minerals.

What is *mud?* What sedimentary rock is derived from mud?

Mud: a mixture of water with particles of silt and clay sizes, along with some sand grains. Lithified mud is *mudstone.*

What is *claystone?*

Claystone: lithified clay, a dense hard rock without natural planes of parting.

What is *shale?* What structure does it have?

Shale: sedimentary rock or mud composition having *fissile* structure, consisting of thin plates and scales that are easily separated.

What are *strata?*

Strata: layers of sedimentary rock, representing variations in composition and texture.

What are *bedding planes?*

Bedding planes: natural planes of separation between strata.

NONCLASTIC SEDIMENTS AND SEDIMENTARY ROCKS

NONCLASTIC SEDIMENTS AND SEDIMENTARY ROCKS 224

Distinguish between the two major divisions of nonclastic sediments.

Two major divisions of nonclastic sediments: (1) *chemical precipitates,* compounds precipitated directly from water; they are *hydrogenic; (2) organically derived sediments,* created by life processes of plants and animals; they are *biogenic.*

What are the *carbonates?* Name three carbonate minerals and give their chemical compositions.

Carbonates: compounds with carbonate (CO_3) in combination with a metal (Ca, Mg). Three common carbonate minerals: *calcite,* $CaCO_3$*; aragonite,* $CaCO_3$; *dolomite,* $CaMg(CO_3)_2$.

To what form of chemical weathering are the carbonate minerals particularly susceptible?

Carbonation: the reaction of carbonic acid (H_2CO_3) with carbonate minerals. Carbonic acid is formed by solution of atmospheric CO_2 in rainwater and subsurface water. Product of carbonation is calcium bicarbonate.

How can carbonate mineral deposition occur?

Reverse process of carbonation allows deposition (precipitation) of crystalline carbonate minerals in soil and rock.

What are the *evaporties?* Name three evaporite minerals and give their chemical composition.

Evaporites: soluble mineral salts precipitated during evaporation of seawater or saline lake water in an arid environment. Three abundant evaporites: *anhydrite,* $CaSO_4$; *gypsum,* $CaSO_4 \cdot 2H_2O$; *halite,* NaCl.

What is *hematite?*

Hematite: a sesquioxide of iron (Fe_2O_3); is major ore of iron.

What is *limestone?* Where does it form?

Limestone: carbonate rock composed largely of calcite ($CaCO_3$); has various origins. *Chalk* variety consists of microskeletons; reef limestone *(reef rock)* is of coral-reef origin.

What is *dolomite?*

Dolomite: rock composed of mineral dolomite, possibly by alteration of marine limestone in contact with seawater.

What is *chert?*

Chert: composed of chalcedony. Occurs as nodules or as solid rock layer called *bedded chert.*

HYDROCARBON COMPOUNDS IN SEDIMENTARY ROCKS

HYDROCARBON COMPOUNDS IN SEDIMENTARY ROCKS 225

What are hydrocarbon compounds? Of what elements are they composed?

Hydrocarbon compounds: complex compounds of hydrogen (H), carbon (C), and oxygen (O). They occur in nature in gaseous, liquid, and solid states; include peat, coal, petroleum, natural gas.

What is *peat?*

Peat: partially decomposed, compacted accumulation of plant remains formed in a bog environment, which may be a freshwater bog or a salt marsh.

What are the *fossil fuels?*

Fossil fuels: collective term for coal, liquid petroleum (crude oil), and natural gas. All originated from organic matter produced by organisms in geologic past.

What is *coal?*

Coal: solid, black hydrocarbon compounds occurring naturally in enclosing sedimentary rock and derived from compacted plant matter buried under thick layers of sediments.

What are *coal seams? coal measures?*

Coal seams are individual coal beds interbedded with mineral strata; *coal measures* are groups of strata containing coal seams.

What is *lignite?*

Lignite: soft brown coal intermediate in composition between peat and true coal.

What is *bituminous coal?*

Bituminous coal: variety of coal found in horizontal strata; it breaks into blocklike masses and may contain identifiable plant remains. Often called "soft coal."

What is *anthracite?*

Anthracite: hard, glasslike coal associated with compressed and folded strata; it consists mostly of fixed carbon.

What is *petroleum?*

Petroleum: Mixture of hydrocarbon compounds occurring as liquid (crude oil), natural gas, or asphalt.

What is *crude oil?*

Crude oil: liquid form of petroleum consisting of a mixture of a large number of hydrocarbon compounds.

What is *heavy oil?*

Heavy oil: crude oil with asphalt base and relatively high density, as distinguished from paraffin-base crude oil of low density.

What is *natural gas?*

Natural gas: mixture of gases of which methane is most abundant, with minor amounts of ethane, propane, and butane.

How does petroleum originate?

Petroleum is thought to have originated as particles of organic matter buried with muds in thick sequences of marine strata.

What steps lead to the accumulation of petroleum?

Heating of sediments converts organic compounds into hydrocarbon compounds; migration of hydrocarbons follows to reach *reservoir rock.*

What is a *reservoir rock?*

Reservoir rock: geologic structure of configuration of strata causing petroleum to be trapped in a reservoir rock.

What is a *reservoir trap?*

Reservoir trap: occurs when reservoir rock is capped or surrounded by dense rock; prevents petroleum from excaping to the surface.

What is *oil shale?*

Oil shale: shale naturally rich in petroleum. Oil shales of Rocky Mountain region are carbonate strata rich in *kerogen,* a waxy hydrocarbon compound capable of conversion into crude oil by heating.

What is *bitumen?*

Bitumen: variety of petroleum that behaves as a solid at low temperatures. Bitumen occurs in sand formations called *bituminous sands (oil sands)* from which petroleum can be extracted by heating.

METAMORPHIC ROCKS

METAMORPHIC ROCKS 226

Compare metamorphic rocks with their original types.

Metamorphic rocks are usually harder, more compact, than parent rocks; have new minerals and structures resulting from kneading action under high pressures.

What is *slate?*

Slate: a compact, fine-grained variety of metamorphic rock, derived from shale and showing well-developed cleavage.

What is *schist?*

Schist: a foliated metamorphic rock in which mica flakes are typically found oriented parallel with foliation surfaces.

What is *foliation?*

Foliation: a type of cleaveage structure in rocks such as slate and shist. In shists, consists of thin and irregularly curved planes of parting.

What is *gneiss?*

Gneiss: variety of metamorphic rock showing banding and commonly rich in quartz and feldspar. Some banded gneisses may have been formed from clastic sedimentary rocks; others (granite gneiss) were derived from granite.

What is *quartzite?*

Quartzite: a metamorphic rock consisting largely of quartz grains, with interstices occupied by silica. Formed from sandstone and siltstone.

What is *marble?*

Marble: a variety of metamorphic rock derived by recrystallization of calcite or dolomite in limestone or dolomite rock.

CYCLE OF ROCK TRANSFORMATION

CYCLE OF ROCK TRANSFORMATION 227

What two environments take part in the cycle of rock transformation?

Both *surface environment* and *deep environment* take part in the *cycle of rock transformation.*

What is the *cycle of rock transformation?* Explain fully. How is the concept of recycling involved?

Cycle of rock transformation: the total circuit of rock changes by which rock of any one class can be transformed into rock of any other class.

Where can we find the original rock of the earth's crust?

Crustal rocks have been continually recycled through geologic time. No rock now in existence is the original rock present at time of origin of earth.

CHAPTER 12—SAMPLE OBJECTIVE TEST QUESTIONS

A. MATCHING

1.	quartz	a. ______	hydrocarbon
2.	amphibole group	b. ______	clay mineral
3.	gabbro	c. ______	silicon dioxide
4.	peridotite	d. ______	halite
5.	granite	e. ______	ultramafic rock
6.	mineral alteration	f. ______	calcium carbonate
7.	illite	g. ______	mafic minerals
8.	calcite	h. ______	basalt
9.	coal	i. ______	hydrolysis
10.	evaporite	j. ______	coarse-grained

B. MULTIPLE CHOICE

1. Plagioclase feldspars are represented by the composition:

   ______a. quartz and mica.

   ______b. aluminosilicate of magnesium and iron.

   ______c. aluminosilicate with sodium, calcium, or both.

   ______d. aluminosilicate of calcium, sodium, and iron.

2. Basalt is an igneous rock

   ______a. commonly occurring as an intrusive body.

   ______b. formed of the same minerals as granite.

   ______c. of the same composition as peridotite.

   ______d. formed of the same minerals as gabbro.

3. The chemically precipitated rocks include

   ______a. limestone, dolomite, and evaporites.

   ______b. limestone, shale, and sandstone.

   ______c. dolomite, siltstone, and shale.

   ______d. dolomite, evaporites, and slate.

C. COMPLETION

1. A silicate mineral containing magnesium and iron, but no aluminum is ______________________.

2. A silicate igneous rock composed mostly of pyroxene, plagioclase feldspar, and sometimes olivine is ______________________.

3. The extrusive equivalent of granite is rhylate.

4. The chemical union of water molecules with minerals is called hydrolisis.

5. Mineral particles detached physically from the parent rock source are described as clastic.

CHAPTER 13

THE LITHOSPHERE AND PLATE TECTONICS

PAGE

THE EARTH'S INTERIOR

THE EARTH'S INTERIOR 228

Name the three major interior zones of the earth.

Three major interior zones of earth: *core, mantle, crust.*

Describe the *core.* Give the radius in km. Of what elements is the core largely composed? What physical property has the outer core? the inner core?

Core: spherical central region of earth, 3500 km (2200 mi) in radius and composed largely or iron. Outer core is liquid; inner core is solid.

Describe the *mantle.* Give thickness in km. Of what material is the mantle composed? What rock at the earth's surface has a similar composition?

Mantle: rock shell, 2895 km (1800 mi) thick, surrounding the core. Mantle is composed of ultramafic silicate rock, or composition similar to periodotite (see Chapter 12).

THE EARTH'S CRUST

THE EARTH'S CRUST 229

What is the *crust?* How thick is the crust? Of what rock type is it largely composed? How does rock type of the upper layer of continental crust differ from its lower layer? Of what rock is the oceanic crust composed?

Crust: outermost earth layer, 5 to 40 km (3 to 25 mi) thick, formed largely of igneous and metamorphic rock. Composition ranges from felsic *granitic rock* in upper layer of continental crust, to mafic basaltic rock in oceanic crust and in lower layer of continental crust.

What is the *Moho?*

Moho: surface of contact between crust and mantle. (Abbreviation of name of seismologist who made the discovery.)

How does thickness of crust beneath continents compare with thickness beneath oceans?

Crust of continents is much thicker than crust beneath oceans Moho under continents lies much deeper than under oceans.

LITHOSPHERE AND ASTHENOSPHERE

LITHOSPHERE AND ASTHENOSPHERE 229

What is the *soft layer* of the mantle? Why is it weak?

Soft layer of mantle is a layer within the upper mantle in which the rock temperature is close to the melting point, causing the rock to be weak. Also called the *asthenosphere.*

What is the *asthenosphere?*

Asthenosphere: the soft, or weak, layer of the mantle; it extends down to about 300 km (185 mi).

What is the *lithosphere?*

Lithosphere: the strong, rigid rock layer above the asthenosphere. Includes both the crust and some of upper part of mantle.

How deep is the asthenosphere beneath the continents?

Beneath continental crust the asthenosphere sets in at an average depth of about 80 km (50 mi); lithosphere is thus about 80 km thick.

How deep is the asthenosphere beneath the oceans?

Beneath the oceanic crust, asthenosphere sets in at about 40 km (25 mi): here the lithosphere is 40 km thick, about half the thickness found beneath the continental crust.

What is important about the physical behavior of the asthenosphere?

The asthenosphere is capable of yielding by slow flowage, permitting the rigid lithosphere to move over it.

What is a *lithospheric plate?*

Lithospheric plate: portion of lithosphere moving as a unit, in contact with adjacent lithospheric plates along contact zones (plate boundaries).

DISTRIBUTION OF CONTINENTS AND OCEAN BASINS

DISTRIBUTION OF CONTINENTS AND OCEAN BASINS 230

What are the first-order relief features of the globe?

Continents and ocean basins are first-order relief features of globe.

As shown on a world map, what percentage of the earth's surface is land, what percent is ocean?

Relative areas as shown on map: land areas 29%; ocean areas 71%.

What submerged features are parts of the continents? At what water depth should the boundary between continents and ocean basins be drawn? What then would be percentages of land and ocean?

Shallow offshore zones (continental shelves) are submerged parts of continents. True limits of continents revealed by boundary drawn at –180 m (–600 ft), giving continents 35% and ocean basins 65%.

SCALE OF THE EARTH'S RELIEF FEATURES

SCALE OF THE EARTH'S RELIEF FEATURES 231

How important are the earth's relief features as compared with a perfectly spherical earth?

Represented to true scale, the earth's greatest relief features (Mt. Everest, ocean trenches) appear as trivial irregularities on a perfect circle.

THE GEOLOGIC TIME SCALE

THE GEOLOGIC TIME SCALE 231

What is *Precambrian time?* How long did it endure?

Precambrian time: all geologic time older than about 570 million years (m.y.). Precambrian time spans about 3 billion years.

What are the *eras* of geologic time? How many are there? Name them in order from oldest to youngest and give ages in millions of years from beginning to end of each.

Eras of geologic time: Three eras in order from oldest to youngest:

Paleozoic Era, –600 to –225 m.y.

Mesozoic Era, –225 to –66 m.y.

Cenozoic Era, younger than –66 m.y.

What are *periods* of geologic time?

Periods of geologic time are subdivisions of the eras, each ranging in duration between 35 and 70 m.y.

What is an *orogeny?*

Orogeny: a major episode of tectonic activity resulting in strata being deformed by crumbling (folding) and breakage (faulting). Eras and most periods ended with an orogeny.

What is an *epoch?*

Epoch: time subdivisions within a period.

What is the *Paleogene Period?*

Paleogene Period: the first three Cenozoic epochs including the Paleocene, Eocene, and Oligocene.

What is the *Neogene Period?*

Neogene Period: includes two Cenozoic epochs, the Miocene and Pliocene together.

What archaic grouping is still used by geologists to subdivide the Cenozoic epochs? (See Table 13.1, pg. 232.)

The *Tertiary Period,* (Paleocene through Pliocene) and the *Quaternary Period* (Pleistocene and Holocene) are archaic groupings.

What method has been used to determine rock ages?

Chemical analyses of radioactive minerals in rocks have yielded rock ages for all eras and periods.

SECOND-ORDER RELIEF FEATURES OF THE CONTINENTS

SECOND-ORDER RELIEF FEATURES OF THE CONTINENTS 232

What two basic subdivisions of the continental masses can be recognized?

Basic subdivisions of continental masses: (1) active belts of mountain-making; (2) inactive regions of old rocks.

In what two ways can mountain ranges be formed?

Mountain ranges can be formed in two ways: (1) *volcanism; (2) tectonic activity.*

What is *volcanism?*

Volcanism: general term for volcano building and related forms of extrusive igneous activity.

What is *tectonic activity?*

Tectonic activity: breaking and bending of the earth's crust and lithosphere in response to internal earth forces.

ALPINE CHAINS

ALPINE CHAINS 233

What is the nature of active mountain-making belts in terms of position and form?

Active mountain-making belts are narrow zones, mostly lying close to continental margins, and consisting of high, rugged mountains.

What are *alpine chains?* Give an example. When were alpine chains formed? What longer features do they make up?

Alpine chains: the active belts of mountain making. Named for European Alps. Alpine mountain chains were formed in Cenozoic Era; comprise arcs linked in chains.

What is a *mountain arc?*

Mountain arc: a curved (arcuate) segment of a mountain chain.

Describe the *circum-Pacific belt.* In what regions does it consist of mountain arcs? Where of island arcs?

Circum-Pacific belt: chain of mountain arcs and island arcs surrounding the Pacific Ocean basin. Consists largely of mountain arcs in North and South America, but of island arcs in western Pacific basin.

What is an *island arc?* Where are they found?

Island arc: a curved (arcuate) line of active volcanic islands. Most occur in western Pacific basin.

What is the *Eurasian-Indonesian belt?* Give the limits of its extent.

Eurasian-Indonesian belt: chain of mountain arcs running from North Africa to Indonesia. Includes Himalaya range.

In terms of world structural regions, to what system do the alpine chains belong?

On world map of structural regions (Figure 13.8, pg. 236) alpine chains belong to the *Alpine system.* (It includes some adjacent inactive belts of orogenic activity in Mesozoic Era.)

CONTINENTAL SHIELDS AND MOUNTAIN ROOTS

CONTINENTAL SHIELDS AND MOUNTAIN ROOTS 233

Name the two structural types within the inactive regions of crust.

Two structural types within inactive regions of crust: (1) shields; (2) roots of older mountain belts.

What are the *continental shields?* Of what rocks are the shields composed? What is the age of most shield rock?

Continental shields: low-lying continental surfaces made up of ancient igneous and metamorphic rocks in a complex arrangement. Most shield rock is of Precambrian age (except for younger sedimentary cover).

Describe the evolution of shields. Through what span of geologic time have they evolved? What is the surface topography of shield areas?

Shields represent regions from which thousands of meters of rock have been eroded during half-billion years of Paleozoic, Mesozoic, and Cenozoic time. Most of shield area is low hills or plateaus.

What are *covered shields?* What is the age of the cover strata?

Covered shields: shield areas bearing a comparatively thin cover of sedimentary strata, usually flat-lying and mostly of Paleozoic and younger age.

What are *exposed shields?* Give an example.

Exposed shields: shield areas lacking in cover; exposing old rocks. Example: Canadian Shield.

What are *epeirogenic movements?*

Epeirogenic movements: slow rising or sinking movements of the crust affecting large areas, but without appreciable breaking (faulting) or bending (folding) of the rocks. A type of movement contrasting with tectonic activity. Movements of shield areas have been largely of the epeirogenic type.

What are *mountain roots?* What is their surface appearance like? Give an example.

Mountain roots: ancient tectonic belts in which folded, faulted, and metamorphic rocks are exposed by long period of erosion. Show as long, narrow ridges and valleys. Example: Appalachian Mountains of North America.

What are the Caledonides?

The Caledonides, a system of mountain roots are remains of Caledonian orogeny at close of Silurian Period (Ireland, Scotland, Scandinavia).

What was the Appalachian orogeny?

Appalachian orogeny at close of Paleozoic Era produced Appalachian chain of mountain roots. (Hercynian orogeny approximately same age in Europe.)

To what system are mountains produced in the Mesozoic Era assigned?

Mountain ranges produced by orogenies of Mesozoic Era (Nevadian and Laramian orogenies) are intermediate in evolution between alpine ranges and ancient mountain roots. Example: Rocky Mountains.

SECOND-ORDER RELIEF FEATURES OF THE OCEAN BASINS

SECOND-ORDER RELIEF FEATURES OF THE OCEAN BASINS 235

What kind of rock makes most of the ocean floor bedrock? What is the age of ocean floor rock?

Ocean floor bedrock almost entirely basalt. Sediment cover present over large areas. Basalt is geologically young, mostly less than 60 million years.

Name the important types of crustal features of the ocean basins.

Crustal features of ocean basins include mid-oceanic ridge, ocean basin floor with abyssal plains, continental rise, continental slope.

What is the *mid-oceanic ridge?* How long is it? Through what ocean basins can it be traced?

Mid-oceanic ridge: a mountainous zone rising to a centrally located axial rift. Can be traced 64,000 km (40,000 mi) through Atlantic, Indian, and Pacific Oceans.

What is the *axial rift?* Where is it found? What crustal activity does its form suggest?

Axial rift: a narrow, trenchlike feature running down the center of the mid-oceanic ridge. Suggests crustal spreading.

What is the *ocean basin floor?* What is the average depth of the floor? What is its position with respect to the mid-oceanic ridge?

Ocean basin floor: broad plains and hill areas averaging about 5 km (17,000 ft) depth and situated between mid-oceanic ridge and continents.

Describe *abyssal plains.* What is the surface form of abyssal plains? What kind of material underlies the plains?

Abyssal plains: extremely smooth, flat areas of ocean basin floor underlain by an accumulation of fine sediment.

THE PASSIVE CONTINENTAL MARGINS

THE PASSIVE CONTINENTAL MARGINS 240

What is the *continental rise?* How is the continental rise related to the continental slope? to the ocean floor? What kind of material underlies the continental rise? Is it thick or thin?

Continental rise: smooth, gradually rising slope leading up from ocean basin floor to foot of continental slope. Underlain by thick sediments brought down from continental slope.

What is the *continental slope?* What lies on the landward side? on the ocean-basin side?

Continental slope: comparatively steep slope descending from brink of continental shelf, flattening out and merging with continental rise.

Describe the *continental shelf.* Describe the slope of the shelf. What is its depth at the outer brink? How wide is it? What kind of material underlies the shelf? How thick is it? What important natural resource may occur within continental shelf deposits?

Continental shelf: shallow, plainlike platform bordering the continents in many places. Surface slopes gently to reach depth of about 180 m (600 ft) at outer brink. Shelf is 120–160 km (75–100 mi) wide off east coast of North America. Thick wedge of sediments underlies continental shelf; possibly rich in petroleum accumulations.

What is the *continental margin?* What features does it include?

Continental margin: narrow zone in which oceanic lithosphere is in contact with continental lithosphere. It includes the shelf, slope, and rise. Oceanic crust and lithosphere underlie the rise; continental crust and lithosphere underlie the shelf and slope.

What are *passive continental margins?* Give an example.

Passive continental margins: margins that have not been subjected to Cenozoic tectonic or volcanic activity; they are not associated with active lithospheric plate boundaries. Example: Atlantic Ocean margins on both east and west sides.

What kind of materials underlie the passive continental margins? (See Fig. 13.12, pg 241.)

Thick sedimentary strata underlie the passive continental margins. Two sediment wedges are recognized: (1) continental-shelf wedge; (2) deep-sea wedge beneath continental rise formed of turbidity current deposits.

What are *turbidity currents?*

Turbidity currents: submarine bottom currents consisting of mud in suspension moving rapidly downgrade, often confined in submarine canyons of the continental shelf and spreading broadly over the continental rise and abyssal plain.

What is a *submarine canyon?* How is it formed?

Submarine canyon: troughlike feature cut deeply into continental slope by turbidity currents; often an extension of a terrestrial river mouth.

What submarine feature usually lies at the foot of a major submarine canyon?

Submarine fan (submarine cone): is built on deep ocean floor at foot of a major submarine canyon. Example: Indus Cone.

THE ACTIVE CONTINENTAL MARGINS 241

What are *active continental margins?*

Active continental margins: continental margins that are marked by active lithospheric plate boundaries, along which tectonic and volcanic activity is present.

What submarine topographic feature is usually associated with an active continental margin?

Active continental margins are associated mostly with *oceanic trenches;* long, narrow, deep troughs bordered on the continental side by a mountain arc or island arc.

What is a *backarc basin?*

Backarc basin: limited area of deep ocean floor underlain by oceanic crust, bounded on the continental side by a deep trench or a passive margin and on the oceanward side by an island arc. Example: Bering Abyssal Plain.

PLATE TECTONICS 241

What is meant by the term *plate tectonics?*

Plate tectonics: general theory of lithospheric plates with their relative motions and boundary interactions.

Define *tectonics.*

Tectonics: study of tectonic activity; the breaking and bending of the entire lithosphere, including the crust.

What is *oceanic lithosphere?*

Oceanic lithosphere: lithosphere underlying the oceanic crust; it is comparatively thin, 50 km (30 mi).

What is *continental lithosphere?*

Continental lithosphere: lithosphere underlying the continental crust; it is comparatively thick, 150 km (95 mi).

Of the two types of lithosphere, which is the denser? Explain.

Oceanic lithosphere, with a basaltic crust, is comparatively dense and stands lower; continental lithosphere, with thick felsic crust, is comparatively buoyant and stands higher.

Describe the geologic activity that takes place along the axis of the mid-oceanic ridge.

Along the axis of mid-oceanic ridge, two oceanic lithospheric plates are spreading apart, while basaltic magma is continually rising in the rift floor and forming new oceanic crust.

What geologic activity is taking place on an active plate boundary between oceanic and continental lithosphere?

Along active plate boundary between oceanic and continental lithosphere, the edge of the oceanic plate is bent down and is descending into the hot, soft asthenosphere, passing beneath the more buoyant continental plate.

What is *subduction?*

Subduction: process of downplunging of one plate (usually oceanic lithosphere) beneath an adjacent plate.

Explain why subduction occurs.

Subduction occurs because the oceanic lithosphere is relatively dense; it sinks under gravity once subduction has started.

What happens to the leading edge of the descending lithospheric plate?

Leading edge of the descending lithospheric plate is heated and softened in contact with the asthenosphere, becoming part of that layer and disappearing.

Does subduction produce magma? Explain.

During subduction the uppermost layer of the descending plate (oceanic crust) is melted and rises as magma, eventually reaching the earth's surface to form a chain of volcanoes parallel with the oceanic trench.

What overall changes are taking place in a single lithospheric plate capped by oceanic crust?

A single plate capped by oceanic crust is simultaneously undergoing *accretion* (growth by addition) at one boundary, and *consumption* (destruction by softening and melting) at the opposite boundary.

PLATE BOUNDARIES

PLATE BOUNDARIES 243

What is a *fault?*

Fault: plane of rock fracture along which there is motion of the rock mass on one side with respect to that on the other.

What is a *transform fault?*

Transform fault: type of fault in which the fracture plane is nearly vertical and motion is horizontal and opposite in crustal rock masses on the two sides of the fault.

Name three kinds of plate boundaries.

Besides *spreading boundaries* and *converging boundaries,* there exist *transform boundaries,* formed by transform faults.

THE GLOBAL SYSTEM OF LITHOSPHERIC PLATES

THE GLOBAL SYSTEM OF LITHOSPHERIC PLATES 243

Describe the global system of lithospheric plates.

Fifteen major lithospheric plates make up the global system; of these, six are very large, the others intermediate or small in size.

Name the six great plates.

Great plates: Pacific, American, Eurasian, African, Austral-Indian, Antarctic.

Name nine lesser plates.

Nine lesser plates are: Nazca, Cocos, Philippine, Caribbean, Arabian, Juan de Fuca, Caroline, Bismark and Scotia.

Which of the great plates consists largely of continental lithosphere?

The Eurasian plate consists largely of continental lithosphere.

What are the general features of the Pacific plate?

Pacific plate: consists almost entirely of oceanic lithosphere. Relative motion is northwesterly.

What are the general features of the American plate?

American plate: includes most of continental lithosphere of Americas as well as oceanic lithosphere west of mid-Atlantic ridge. Eastern boundary: spreading; western boundary: converging with subduction.

What is a *subplate?*

Subplate: plate of secondary importance lying within a great plate and set apart for recognition because it seems to have an independent form of tectonic activity. At some future time it may separate from the parent plate.

SUBDUCTION TECTONICS AND VOLCANIC ARCS

SUBDUCTION TECTONICS AND VOLCANIC ARCS 246

What geologic activities are taking place along the trench axis of an active continental margin?

During subduction, the trench axis receives sediment from two sources: (1) ocean-floor sediment carried on moving oceanic plate; (2) terrestrial (land-derived) sediment swept into deep water by currents.

What happens to sediment in the trench floor?

Sediment mass in trench floor is deformed into wedges to form an *accretionary prism.*

What happens to the sediments of the accretionary prism?

Accretionary prism is changed into metamorphic rock and added to continental crust.

What is a *tectonic crest?* How does it form?

Rising edge of the accretionary prism forms a ridgelike *tectonic crest;* it may appear as an island chain called a *tectonic arc.*

What feature lies between the tectonic arc and the mainland?

A *forearc trough* (shallow sea) forms between the tectonic arc and the mainland; it traps large amounts of terrestrial sediment.

Define *forearc trough.*

Forearc trough: long, narrow submarine trough lying between a tectonic arc and a volcanic arc and receiving sediments from either or both arcs.

Explain how volcanism is tied in with subduction.

Basaltic magma, derived by melting of upper crustal layer of descending plate, is changed to andesite magma at base of crust; it rises through crust to form andesitic volcanoes and lava flows.

OROGENY

OROGENY 247

Describe the internal geologic structure of alpine mountains.

Alpine mountain chains contain strongly folded and faulted strata of marine origin. *Folds* are overturned (recumbent) and broken by *overthrust faults* to form *thrust sheets,* the total deformed mass being an *orogen.*

Define *folds* of strata.

Folds: wavelike structures (corrugations) of strata resulting from compressional movements of crust.

What is an *overthrust fault?*

Overthrust fault: fault with its plane of sliding gently inclined from the horizontal on which one rock slice rides over the other, often for many kilometers.

What is a *thrust sheet?*

Thrust sheet: rock mass on upper side of overthrust fault (called *nappes* in the European Alps).

Describe *gravity gliding.*

Gravity gliding: the phenomena of thrust sheets sliding "downhill" under the force of gravity.

OROGENS AND COLLISIONS

OROGENS AND COLLISIONS 248

What two types of orogens can be recognized?

Two types of orogens: (1) resulting from impact of small lithospheric mass against a full-sized mass of continental lithosphere; (2) from collision of two very large bodies of continental lithosphere (full-sized plates) uniting them permanently.

What kind of small crustal mass might impact a large plate?

Commonly, the impacting mass is a volcanic island arc, which is narrow (but may be of great length).

What other kinds of crustal masses might be involved?

Microcontinents: small continental fragments rifted away from a large continent, may also be involved in collision.

Describe the *arc-continent collision.* How can the two bodies move toward each other?

Arc-continent collision: collision of a volcanic island arc with a large continent. A subduction zone separates the two bodies, allowing the ocean between them to close up.

What activities are occurring along the subduction zone?

Along the subduction zone an accretionary prism is growing larger and thicker.

What happens next?

Next, the accretionary prism meets the passive continental boundary, impacting the strata of the passive margin and forcing them into folds and low-angle overthrusts (process called "telescoping").

What is a *décollement?*

Décollement: detachment and extensive sliding of a rock layer, usually sedimentary, over a near-horizontal basal rock surface.

What happens in the final stage?

In the final stage of collision, rocks of the accretionary prism and the volcanic arc itself are crumpled into metamorphic rock and intruded by batholiths.

ACCRETED TERRANES OF WESTERN NORTH AMERICA

ACCRETED TERRANES OF WESTERN NORTH AMERICA 250

What if a *terrane?*

Terrane: continental crustal rock unit having a distinctive set of lithologic properties, reflecting its geologic history, that distinguish it from adjacent or surrounding continental crust.

Where are terranes to be found?

Numerous terranes form a broad zone of western North America, extending from Alaska to Mexico. At least 50 different terranes have been identified.

What feature forms the boundary between terranes?

Typically, adjacent terranes are separated by a fault contact; commonly the fault is of the transcurrent type.

Where did these terranes originate?

Some terranes may have traveled distances as great as several thousand kilometers to make collision with the North American continent (evidence from paleomagnetism).

CONTINENT-CONTINENT COLLISIONS

CONTINENT-CONTINENT COLLISIONS 251

What is a *continent-continent collision?*

Continent-continent collision: collision between two large masses of continental lithosphere along a subduction plate boundary, resulting in a continental suture.

What is a *continental suture?*

Continental suture: orogen formed by continent-continent collision.

Where has such a collision taken place in comparatively recent geologic time?

A collision boundary, or suture, lies along the southern margin of the Eurasian plate.

Which plates impacted Eurasia?

African, Persian, and Austral-Indian plates collided with the Eurasian plate in Cenozoic time, closing the intervening ocean.

Which portion of the collision boundary is still active?

The Himalayan arc is presently undergoing continued tectonic activity in late stages of collision.

CONTINENTAL RUPTURE AND NEW OCEAN BASINS

CONTINENTAL RUPTURE AND NEW OCEAN BASINS 253

What is *continental rupture*?

Continental rupture: splitting apart or rifting apart of a single plate formed of continental lithosphere.

What are *fault-block mountains?*

Fault-block mountains: mountains formed when the first crust is lifted and stretched apart as the lithospheric plate is arched upward; result of extensional tectonics.

What geologic features develop during continental rupture?

Continental rupture initially produces a *rift valley* (long narrow valley with steep walls), followed by development of numerous fault blocks. A narrow ocean then appears, becoming wider. New oceanic crust and lithosphere forms in the floor of the widening ocean basin (example: Red Sea).

Describe the development of the spreading plate boundary.

Spreading plate boundary (axial rift) develops offsets connected by transform faults. These produce *transform scars (fracture zones)* extending far across ocean floor.

THE BREAKUP OF PANGEA

Describe the hypothesis of continental drift. What was *Pangaea?* Describe the breakup of Pangaea. How was the North Atlantic Ocean basin formed?

What is *continental drift?* To what extent has this program been accepted by scientists today?

THE BREAKUP OF PANGEA 254

Hypothesis of continental drift, proposed by A. Wegener many decades ago, has a single supercontinet, *Pangaea,* break up and form pieces drifting apart to become present continents and intervening ocean basins.

Continental drift: early concept of continents slowly separating from one another after breakup of a single parent continent. General program of separation is now generally accepted and has been fitted into plate tectonics.

MODELING THE EARTH'S INTERNAL ENERGY SYSTEM

What kind of energy system is represented by lithospheric plate motions?

What if the speculation on future plate motions and tectonic activity?

MODELING THE EARTH'S INTERNAL ENERGY SYSTEM 255

The plate tectonic system is an *exponential decay system* of energy flow, powered by an initial store of nuclear energy.

It is speculated that plate motion and tectonic activity will decrease as eons of time pass because the earth's stores of radioactive elements is diminishing, and the energy for a moving plate system is decreasing.

CHAPTER 13—SAMPLE OBJECTIVE TEST QUESTIONS

A. MATCHING

1.	core	a. _______	soft layer
2.	Moho	b. _______	Precambrian age
3.	Paleozoic	c. _______	abyssal plain
4.	fault	d. _______	liquid iron
5.	shields	e. _______	era
6.	ocean basin floor	f. _______	continental drift
7.	asthenosphere	g. _______	overthrust
8.	Pangaea	h. _______	base of crust

B. MULTIPLE CHOICE

1. The Paleozoic Era is

 ________a. younger than the Mesozoic Era.

 ________b. younger than the Precambrian time.

 ________c. about 2 Billion years old.

 ________d. preceded by the Cenozoic Era.

2. Rifting of the crust

 ________a. indicates crustal spreading.

 ________b. is accompanied by folding.

 ________c. is typical of the zones of alpine structure.

 ________d. causes overthrust faulting.

3. The continental shelf

 ________a. is a region of exposed shield rock.

 ________b. has depths mostly greater than 5000 m.

 ________c. is underlain by a thick deposit of sediments.

 ________d. slopes very steeply toward the ocean floor.

C. COMPLETION

1. Name the three major interior zones of the earth: ______________________________

 ______________________________ and ______________________________,

2. Motion of a lithospheric plate takes place by flowage within the soft layer, or ____________

 ______________________ beneath.

3. Down-plunging of one edge of a lithospheric plate is called ______________________________

 ______________________.

4. Two lithospheric plates may slide past one another along a ______________________________ fault.

5. The number of major lithospheric plates is __________, of these the ______________________ ____ plate is the largest one entirely within oceanic crust.

6. A narrow trench-like feature running down the center of the mid-oceanic ridge is known as the ____________________________________.

CHAPTER 14

VOLCANIC AND TECTONIC LANDFORMS

PAGE

What are *landforms?* Give examples.

Landforms: distinctive configurations of the land surface produced by natural processes. Examples: hill, plain, sea cliff.

What is *geomorphology?*

Geomorphology: the science of landforms, including their history and processes of origin.

Define *denudation.*

Denudation: the lowering of the continental sufaces by removal and transportation of mineral matter through the action of running water, waves and currents, glacial ice, and wind.

INITIAL AND SEQUENTIAL LANDFORMS — 258

What are *initial landforms?* Give two examples.

Initial landforms: landforms produced directly by volcanic and tectonic activity. Examples: volcano, fault block.

What are *sequential landforms?* Give two examples.

Sequential landforms: landforms shaped by processes and agents of denudation, namely, weathering and mass wasting, running water, glaciers, waves, and wind. Examples: floodplain, canyon, sea cliff.

What is a *landmass?*

Landmass: large area of continental crust lying above sea level.

What concept of landscape relates internal earth forces to external earth forces?

A landscape represents a stage in the contest between internal earth forces and external earth forces, as the former cause the crust to be elevated and the latter work to lower the land to sea level.

VOLCANIC ACTIVITY — 259

What is a volcano? Of what materials is a volcano built?

Volcano: a conical or dome-shaped structure built by emission of lava flows and accumulations of tephra.

What gives form and dimension to a volcano?

Form and dimension of a volcano is determined by (1) type of lava; (2) presence or absence of tephra; (3) type of magma.

What kind of lava gives explosive eruptions? quiet eruptions?

Explosive behavior of volcanoes is typical of felsic lavas (rhyolite, andesite); quiet eruptions typical of mafic magma (basalt).

COMPOSITE VOLCANOES 259

What is a volcanic *crater?*

Crater: the central summit depression associated with the principal vent of a volcano.

What is a *composite volcano?* Give two examples.

Composite volcano: high, steep-sided cone built of lava flows and ash layers. Most are built of felsic lavas; show explosive behavior. Examples: Mount Hood, Fujiyama, Mount Mayon.

What is a *nuée ardente?*

Nuée ardente: glowing volcanic avalanche; a cloud of incandescent dust and gas; it travels very rapidly down the steep side of a volcano, destroying all life in its path.

How are composite volcanoes related to plate tectonics?

Chains of composite volcanoes lie mostly above active subduction zones, such as Aleutian arc, Andes arc, Sumatra-Java arc.

CALDERAS 260

What is a *caldera?* What happens to the missing body of rock?

Caldera: large, saucer-shaped depression formed by explosive destruction of a composite volcano. Most of the mass of the upper cone subsides; some is blown out as tephra.

Give examples of caldera explosions of recent times.

Krakatoa (1883) and Katmai (1912) are examples of volcanic explosions producing new calderas.

Give an example of an ancient, prehistoric caldera event.

Crater Lake, Oregon, is caldera produced about 6600 years ago, when former Mount Mazama was destroyed.

FLOOD BASALTS AND SHIELD VOLCANOES 260

What is a *mantle plume?*

Mantle plume: slowly rising column of heated mantle rock postulated to occur from place to place in the asthenosphere.

What is a *hot spot?* How is it related to volcanic activity?

Hot spot: center of intrusive igneous and volcanic activity thought to be located over a rising mantle plume. Magma of basaltic composition can emerge from hot spot after making its way up through the continental crust.

What are *flood basalts?* How thick are they? Give an example.

Flood basalts: thick accumulations of highly fluid basalt lava flows. Total accumulation may be thousands of meters thick. Example: Columbia Plateau.

Describe a *shield volcano.* Of what rock is it built?

Shield volcano: broad-topped volcano with low side slopes, built of quietly erupting basalt flows.

Describe the Hawaiian Islands as examples of shield volcanoes. Give height above ocean floor. What surface features do they have? Where do flows emerge?

Hawaiian Islands are major examples of shield volcanoes, built upward from deep ocean floor to summit altitudes of 4000 m (13,000 ft). Have steep-sided central depression at summit (form of caldera); many pit craters.

How is the Hawaiian island chain related to plate motion?

Hawaiian volcanic chain is interpreted as result of Pacific lithospheric plate moving over a mantle plume.

In what other type of geologic situation do shield volcanoes and flood basalts occur?

Iceland, an island of basaltic lava flows and shield volcanoes, is situated over the mid-oceanic ridge of crustal spreading; it is on a spreading plate boundary.

CINDER CONES

CINDER CONES 263

Describe a *cinder cone.*

Cinder cone: conical or dome-shaped hill built of coarse basaltic tephra ejected from a narrow vent; a form of volcano. Cinder cones form in groups. Basalt lava flows may be present in same region.

ENVIRONMENTAL ASPECTS OF VOLCANIC ACTIVITY AND LANDFORMS

ENVIRONMENTAL ASPECTS OF VOLCANIC ACTIVITY AND LANDFORMS 263

Describe volcanic activity as an environmental hazard. What causes destruction? Give example. What ocean phenomenon is sometimes generated by volcanic explosion?

Volcanic eruptions are severe environmental hazard. Destruction occurs through clouds of incandescent gases (example, Pelée), fall of tephra, lava flows, mudflows, and earthquakes. Seismic sea waves a hazard on coasts.

Describe the environmental aspects of landforms of volcanoes and lava flows. Of what value to humans are volcanoes? Name two volcanoes protected by inclusion in National Parks.

Environmental aspects of landforms of volcanoes and lava flows: lava surfaces remain rough and barren until weathering produces soil. Volcanoes are valuable resources for recreation and tourism. National parks protect many volcanoes. Examples: Mt. Rainier, Crater Lake, Mauna Loa.

GEOTHERMAL ENERGY RESOURCES 264

What is *geothermal energy?*

Geothermal energy: heat energy originating within the earth's crust.

What is the *geothermal gradient?*

Downward rate of increase in rock temperature in the crust is the *geothermal gradient:* it averages 20 to 40 C° per km.

What are *geothermal localities?*

Heated rock lying close to the surface and available as an energy source is described as a *geothermal locality.*

What are *hot springs?*

Hot springs: places at which rising hot water (close to boiling point) reaches the earth's surface to form natural springs.

What is a *geyser?*

Geyser: jetlike emission of steam and hot water at a geothermal locality.

What is a *fumarole?*

Fumarole: jet of superheated steam emerging from ground at a geothermal locality.

Describe deep sources of hot water.

Hot ground water occurs at great depth in certain localities and can be brought to the surface through drilled wells. Example: Imperial Valley.

How can dry hot rock be an energy source?

In certain localities at a depth of 2 to 5 km, solid igneous rock of a batholith is still hot. Water forced down to the hot rock in drilled wells and returned to surface is highly heated.

What are *geopressurized zones?*

At certain places deep in continental shelf strata hot water is present under high pressure with natural gas. Capability exists for using this heat as energy source.

LANDFORMS OF TECTONIC ACTIVITY

LANDFORMS OF TECTONIC ACTIVITY 266

Distinguish between tectonic activity involving compression and that involving rifting.

Landforms of tectonic activity involve compression and folding (in subduction zones and continental collision), and rifting (pulling apart) to produce fault blocks.

What is a *fault block?*

Fault block: blocklike rock mass lying between two parallel *normal faults.*

FORELAND FOLD BELTS

FORELAND FOLD BELTS 266

What is meant by *folding* of strata?

Folding: the process of formation of wavelike undulations, or *folds,* in strata or other layered rocks, under compressional forces.

Under what geologic circumstances has folding occurred on a large scale?

Large-scale folding of strata has occurred on passive continental margins impacted during continental collisions.

What is an *anticline?*

Anticline: archlike upfold in strata or other layered rocks.

What is a *syncline?*

Syncline: troughlike downfold in strata.

What are *foreland folds?*

Foreland folds: folds produced on the continental foreland by arc-continent and continent-continent collisions.

Cite two examples of foreland folds produced in Cenozoic time

Examples: Jura Mountains of the Alps; Zagros Mountains of Iran.

FAULTS AND FAULT LANDFORMS

FAULTS AND FAULT LANDFORMS 266

Define fault. How can a fault be recognized?

Fault: a fracture surface along which slippage movement has occurred, causing displacement of one rock mass with respect to the other.

What is a *fault plane?*

Fault plane: surface of slippage between two fault blocks moving relative to each other during faulting.

What is a *fault line?*

Fault line: the surface trace of a fault.

What is a *normal fault?*

Normal fault: fault in which fault plane has steep angle and the downthrown block occupies the upper side of the fault plane.

Describe a *fault scarp.* What kind of fault has a fault scarp? How high is a fault scarp? How long?

Fault scarp: a clifflike feature produced by movement on a fault, usually a normal fault. Fault scarps range in height from a few meters to many hundreds of meters, in length up to many km.

Define *graben.*

Graben: trenchlike depression representing the surface of a fault block dropped down between two opposed normal faults.

Define *horst.*

Horst: a fault block uplifted between two normal faults.

What is a *reverse fault?*

Reverse fault: type of fault in which one fault block rides up over the other on a steep fault plane.

What is a *transcurrent fault?* What surface features does it show?

Transcurrent fault: fault in which dominant motion is horizontal, in a direction parallel with the fault plane. On a flat plain no scarp would result. Narrow trench, or rift, commonly marks fault line.

What is a *low-angle overthrust fault?*

Low-angle overthrust fault: fault involving largely horizontal motion of a thrust sheet on a thrust plane.

THE RIFT VALLEY SYSTEM OF EAST AFRICA

THE RIFT VALLEY SYSTEM OF EAST AFRICA 268

What is a *rift valley?*

Rift valley: trenchlike valley with steep, parallel sidewalls; essentially a graben between two normal faults.

Under what geologic circumstances do rift valleys form?

Rift valleys appear during continental rifting, in an early stage prior to the opening of a new ocean basin.

What other activity besides faulting is typical of rift valleys?

Volcanic activity, including extrusion of basaltic lavas and the building of volcanic cones, typically accompanies rifting.

What future geologic events can be forecast for the East African rift-valley system?

In the future, that part of Africa east of the rift-valley system may separate from the continent to form a new lithospheric plate, moving northeastward.

EARTHQUAKES

EARTHQUAKES 270

What is an *earthquake?* What kind of waves are involved?

Earthquake: an oscillating motion of the ground surface produced by the passage of a train of *seismic waves.*

Where is earthquake energy released?

Earthquake energy is released during slippage on an active fault from a point called the *earthquake focus.*

Define *seismic waves.* In what direction do they travel?

Seismic waves: earth waves generated principally by faulting, and traveling outward from point of energy release in widening circles.

Describe the San Andreas Fault. What type of fault is it? How long is it? What features indicate the direction of movement?

San Andreas Fault, California, an example of an active earthquake-producing fault; a transcurrent fault 1000 km (600 mi) long. Slippage causes offsetting of linear features crossing the fault line (roads, fences).

What is a *fault creep?*

Fault creep: slow, steady displacement of a fault that tends to reduce stored energy accumulation.

EARTHQUAKE ENERGY— THE RICHTER SCALE

EARTHQUAKE ENERGY— THE RICHTER SCALE 270

What is the *earthquake magnitude (M)?*

Earthquake magnitude (M): the quantity of energy released during an earthquake at the place where fault slippage originates.

What is the *focus*?

Focus: initial point of rock eruption.

What is the *epicenter?*

Epicenter: the point on the surface directly above the focus.

What is the *Richter scale?*

Richter scale: a rating system of earthquake magnitude or energy release. Numbers range from 0–8.5.

EARTHQUAKES AND PLATE TECTONICS

EARTHQUAKES AND PLATE TECTONICS 272

What is the relationship between earthquakes and lithospheric plates? Where is the greatest seismic activity found?

Earthquakes are concentrated near the boundaries of lithospheric plates. The greatest seismic acitivity is found where oceanic plates are undergoing subduction.

What other classes of boundaries are likely to be seismically active?

Transform boundaries, such as the San Andreas Fault and spreading boundaries, such as mid-oceanic ridges, are most likely to be seismically active.

EARTHQUAKES AS ENVIRONMENTAL HAZARDS 274

EARTHQUAKE INTENSITY SCALES

What does an *earthquake intensity scale* measure?

Which scale is used in the U.S.?

Describe the features of the modified Mercalli scale. (See Table 14.2, pg. 274.)

What are *isoseismals?*

EARTHQUAKE INTENSITY SCALES 274

Earthquake intensity scale measures observed earth-shaking effects from an earthquake.

The scale used in the U.S. is called the *modified Mercalli scale.*

The scale: (1) recognizes 12 levels of intensity; (2) each intensity describes phenomena people experience during the earthquake.

Isoseismals: the numbered lines show concentric zones of equal intensity gathered from reports after an earthquake.

EARTHQUAKE GROUND MOTIONS

What instrument is used to measure ground motions during earthquakes?

What is peak acceleration?

What is the secondary effect of a severe earthquake?

EARTHQUAKE GROUND MOTIONS 275

The motion seismograph measures earthquake ground motion.

Peak acceleration: peak rate of change of velocity. Describes severity of ground shaking in an earthquake.

Secondary effect of a severe earthquake is the collapse of the ground under the force of gravity because ground shaking reduces the strength of the earth material on which heavy structures rest.

SEISMIC SEA WAVES

What is a *seismic sea wave (tsunami)?* How long are the waves? How do they cause damage?

SEISMIC SEA WAVES 276

Seismic sea wave (tsunami): train of ocean waves generated by submarine earthquake (or other earth movement) and propagated across ocean in widening circles. Waves are long, low; cause rise of ocean level on distant shores. Highly destructive surf results.

ASSESSING NATIONAL EARTHQUAKE HAZARDS

How do geologists map areas where earthquake damage risk is high?

ASSESSING NATIONAL EARTHQUAKE HAZARDS 277

Geologists map areas where earthquake damage is high by intensive field study of geologic evidence of recent seismic activity.

What major features are shown on the U. S. risk map of maximum intensity of ground shaking?

The central and eastern portions of the U. S. show moderate earthquake risk. Western portions show moderate to high risk, especially near visible fault systems.

How does the Nuclear Regulatory Commission define *capable fault?*

Capable fault: a fault that has exhibited movement at or near the ground surface at least once within the past 35,000 years or movement of a recurring nature within the past 500,000 years.

EARTHQUAKES AND URBAN PLANNING—THE SAN FERNANDO EARTHQUAKE

EARTHQUAKES AND URBAN PLANNING—THE SAN FERNANDO EARTHQUAKE 278

Describe the forms of destruction and damage accompanying the San Fernando earthquake.

San Fernando earthquake, 1971, was brief but produced extremely severe ground shaking. Effects included collapse of some large buildings, damage to water storage dam and power transmission station, and freeway collapse and cracking.

What recommendation did the National Academy of Sciences offer?

National Academy of Sciences panel reported that building codes were not adequate, many changes and improvements are needed. Major earthquake on nearby San Andreas Fault could bring vastly greater destruction.

HAZARDS OF THE SAN ANDREAS FAULT ZONE

HAZARDS OF THE SAN ANDREAS FAULT ZONE 279

What are the three major areas of the San Andreas Fault that are presently locked?

The San Andreas Fault is presently locked in the San Francisco, Los Angeles, and Imperial Valley regions of California.

In which area is an earthquake considered most imminent?

An earthquake is considered most imminent in the Imperial Valley section.

THE LOMA PRIETA EARTHQUAKE OF 1989 280

Near what California city was the Loma Prieta earthquake centered?

The Loma Prieta earthquake was centered near Santa Cruz.

Where was damage in San Francisco concentrated? Why?

Earthquake damage was centered on areas underlain by water-saturated sand and mud. Under the influence of shaking, these materials liquified, losing structural strength and amplifying ground motions.

What are the two recurring themes to be learned for the Loma Prieta earthquake?

The two recurring themes to be learned from the Loma Prieta earthquake are (1) local geologic conditions strongly influence damage; and (2) the patterns of shaking repeats that of earlier earthquakes.

CHAPTER 14—SAMPLE TEST QUESTIONS

A. MATCHING

1. caldera
2. cinder cone
3. graben
4. earthquake
5. tsunami

a. _______ explosion
b. _______ horst
c. _______ tephra
d. _______ focus
e. _______ sea wave

B. MULTIPLE CHOICE

1. A hot spot is

_______ a. the source of basaltic magma for shield volcanoes.
_______ b. the source of andesite magma for composite volcanoes.
_______ c. situated above a mantle plume.
_______ d. (both a. and c. above).

2. In a normal fault, the direction of displacement is

_______ a. largely vertical.
_______ b. largely horizontal.
_______ c. the same as in a reverse fault.
_______ d. the same as in a low-angle overtthrust fault.

3. The Richter scale

_______ a. measures earthquake frequency.
_______ b. predicts earthquakes.
_______ c. measures earthquake intensity.
_______ d. tells distance to the earthquake focus.

C. COMPLETION

1. A tall, steep-sided conical volcano composed of felsic lavas and tephra is a _______________ _______________ volcano.
2. Basaltic volcanoes of dome shape with gentle side slopes are called _______________ _______________ volcanoes.
3. A small volcano formed entirely of basaltic tephra is a _______________ _______________.
4. A fault on which the movement is largely horizontal on a nearly vertical fault plane is a _____ _______________ fault.

CHAPTER 15

LANDFORMS OF WEATHERING AND MASS WASTING

PAGE

WEATHERING AND MASS WASTING

WEATHERING AND MASS WASTING 281

Define *weathering.*

Weathering: processes by which rock is physically disintegrated and chemically decomposed during exposure to atmospheric influences.

Define *regolith.*

Regolith: the products of rock weathering.

Define *bedrock.*

Bedrock: solid rock unaltered by weathering process.

Define *sediment.*

Sediment: detached mineral particles deposited and transported by fluid medium (air, water, glacial ice).

Define *mass wasting.* Does mass wasting include the action of the fluid agents?

Mass wasting: the spontaneous downward movement of soil, regolith, and rock under the influence of gravity. Mass wasting does not include action of the fluid agents.

THE WASTING OF SLOPES

THE WASTING OF SLOPES 282

What is meant by *slope?* What role do slopes play in geomorphology?

Slope: any portion or element of the land surface; often with reference to hillslopes, which are the inclined ground surfaces extending down from divides and summits to valley floors.

What earth materials are usually seen on a hillslope?

On a typical hillslope, soil and *residual regolith* cover the bedrock, except for some *outcrops.*

What is *residual regolith?*

Residual regolith: regolith derived directly by weathering of the underlying bedrock.

What is an *outcrop?*

Outcrop: surface exposure of undisturbed bedrock showing little internal alteration by weathering processes.

What is *transported regolith?*

Transported regolith: regolith formed of mineral matter carried by fluid agents from a distant source and deposited upon the bedrock or upon older regolith. Examples alluvium, beach sand, dune sand.

PHYSICAL WEATHERING PROCESSES AND FORMS

What is *physical weathering?*

PHYSICAL WEATHERING PROCESSES AND FORMS 282

Physical weathering: physical processes that convert hard, massive bedrock into fragments ranging in size from boulder to silt.

Frost Action

What is *frost action?* Where is frost action most effective in its action?

What are *joints?*

Define *block separation.*

Define *granular disintegration.*

What is a *felsenmeer?*

Frost Action 282

Frost action: rock breakup by forces accompanying the freezing of water. Limited to midlatitude and high-latitude climates and cold alpine climates at high altitude.

Joints: a system of fractures that cut through bedrock.

Block separation: separation of individual joint blocks during the process of physical weathering.

Granular disintegration: grain-by-grain breakup of the outer surface of coarse-grained rock, yielding gravel and leaving behind rounded boulders.

Felsenmeer: expanse of large blocks of rock produced by joint block separation and shattering by frost action at high altitudes or in high latitudes. Word is German, means "rock sea."

Salt-Crystal Growth

Explain how *salt crystal growth* takes place in rock to cause disintegration. Under what climate is the process most effective? What features are produced by this process?

Salt Crystal Growth 284

Salt crystal growth takes place in pores of exposed rock as capillary water films evaporate in dry weather. Crystal growth forces grain-by-grain breakup of rocks. Found extensively in dry climates; produces niches and shallow caves in sandstone cliffs. Also affects masonry surfaces. Deicing salts cause similar breakup of highway pavement.

Wedging by Plant Roots

How do plant roots accomplish physical weathering?

Wedging of Plant Roots 284

Plant roots create pressure by growth of tiny rootlets which loosen rock particles.

Temperature Change As a Cause of Rock Disruption

How does temperature change disrupt rocks?

Temperature Change As A Cause of Rock Disruption 282

Temperature change creates expansion and contraction of mineral grains, which can break rocks apart if continues for long periods.

Sheeting Structure and Exfoliation Domes

What is *unloading?* How is denudation involved in unloading? What form of disintegration accompanies unloading?

Sheeting Structure and Exfoliation Domes 284

Unloading: the relief from confining pressure as deep-seated rock is gradually brought to the surface by denudation. Unloading is accompanied by rock expansion, which causes shells and sheets to break free.

What is *sheeting structure?* In what material does it develop? What human activity results in sheeting structure?

Sheeting structure: sheets of solid bedrock separated from the parent mass through expansion accompanying unloading. Seen in rock quarries; facilitates removal of rock slabs.

What is an *exfoliation dome?* What form does it take? How is it caused? Give an example.

Exfoliation dome: a smoothly rounded rock knob or hilltop bearing rock sheets or shells produced by unloading. Example: Yosemite domes.

CHEMICAL WEATHERING PROCESSES AND FORMS

CHEMICAL WEATHERING PROCESSES AND FORMS 285

What are the dominant processes of chemical weathering?

The dominant processes are (1) oxidation; (2) carbonic acid action, and; (3) hydrolysis.

Spheroidal Weathering and Saprolite

Spheroidal Weathering And Saprolite 286

What is *spheroidal weathering?* What forms does it produce? Where does the weathering take place? What kind of rock is most strongly affected by spheroidal weathering?

Spheroidal weathering: production of thin, soft concentric shells of decomposed rock as chemical weathering penetrates joint blocks under a cover of regolith. Hydrolysis is a major cause of spheroidal weathering in igneous rocks.

What is *saprolite?* How is it formed? In what kinds of rocks? What processes are involved? In what climates is saprolite most commonly found? What engineering problems are caused by the presence of saprolite? Explain.

Saprolite: thick residual regolith formed by decomposition of igneous and metamorphic rocks of silicate mineral composition. Minerals are softened by oxidation and hydrolysis. Commonly seen as regolith in warm, moist climates. Plastic clay minerals in saprolite yield under load of heavy human-made structures.

What is *grus?*

Grus: loose gravel formed by granular disintegration of coarse-grained felsic igneous rock, such as granite, in a dry climate.

Effects of Carbonic Acid Action

Effects of Carbonic Acid Action 286

What kinds of rock are most susceptible to the action of carbonic acid?

Carbonic acid action affects carbonate sedimentary rocks (limestone, dolomite) and marble.

What surface features are produced by carbonic acid action?

Carbonic acid action upon carbonate rocks creates minor surface features such as cupping, rilling, or grooving.

In what way is carbonic acid action important in denudation?

Carbonic acid action is major agent of denudation in limestone regions in humid climates. Land surface is lowered rapidly to form valleys.

MASS WASTING

MASS WASTING 287

What concept ties gravity with downward motion of earth materials?

All bedrock, regolith, and soil is constantly under force of gravity and will move downward when internal strength falls below limit of support.

Define *mass wasting.*

Mass wasting: spontaneous downward movement of soil, regolith and bedrock under the influence of gravity.

What range of rates of motion is spanned by the various forms of mass wasting?

Mass wasting ranges in rate from imperceptible creep motion of soil and regolith to high-speed fall of enormous rock masses.

Rockfall and Talus Cones

Rockfall and Talus Cones 288

What is *rockfall?*

Rockfall: free fall of masses of bedrock from a steep cliff face.

What is a *talus slope (scree slope)?*

Talus slope (scree slope): slope formed of *talus,* an accumulation of loose rock fragments derived by fall of rock from a cliff.

What is a *talus cone?*

Talus cone: accumulation of talus in the form of a partial cone with apex at top, heading in a ravine or gully. Coarsest particles (boulders) are at base; finest at apex.

What is the *angle of repose?*

Angle or repose: natural angle held by a slope composed of loose particles of sand grade or coarser sizes.

Talus Creep and Rock Glaciers

Talus Creep and Rock Galciers 289

Define *rock creep.*

Rock creep: the slow down-slope motion of rock fragments of the talus cones.

What are *rock glaciers?*

Rock glaciers: tonguelike bodies of talus fragments produced by creep.

Soil Creep

Soil Creep 289

What is *soil creep?*

Soil Creep: slow downhill movement of soil and regolith.

What causes soil creep?

Soil creep is caused by (1) heating and cooling of soil; (2) growth of frost needles; (3) alternate drying and wetting of soil; (4) animal trampling and burrowing; (5) earthquake shaking.

Earthflows

Earthflows 290

Define *earthflow.* What materials are involved in earthflow? Is the movement rapid or slow? What forms are produced?

Earthflow: moderately rapid downhill flowage of masses of water-saturated soil, regolith, or weak shale. Forms steplike terrace at top, bulging toe at base.

In what way are earthflows induced or aggravated by human activities? Give an example.

Earthflows are induced or aggravated by humans through waste water infiltrating the ground in residential areas. Example: Portuguese Bend "landslide", Palos Verdes Hills, southern California.

Earthflows Involving Quick Clays

Earthflows Involving Quick Clays 290

Describe earthflows in layered clays and silts of St. Lawrence River terraces.

In layered silts and clays of Pleistocene terraces in St. Lawrence lowland and elsewhere, earthflows are horizontal motions of large, flat platelike masses, moving into river channel; motion involves *quick clays.*

What are *quick clays?*

Quick clays: clay layers that undergo *spontaneous liquefaction* when disturbed.

What is *spontaneous liquefaction?*

Spontaneous liquefaction: the sudden change of state of water-saturated solid clays into near-liquid state when arrangement of colloidal particles is disturbed.

Mudflow

Mudflow 291

What is a *mudflow?* What causes mudflows? Where do they originate? In what climate are mudflows common? How are mudflows produced during volcano eruptions?

Mudflow: rapid flowage of mud stream down canyon floors and spreading out upon piedmont surfaces. Produced by heavy rains or snowmelt upon mountain slopes in arid regions. Leaves sheets of bouldery debris. Mudflows also occur in volcanic ash accompanying volcano eruptions.

Debris Floods and Avalanches

Debris Floods and Avalanches 292

What is a *debris flood?* Compare it with a mudflow. Where are debris floods an important environmental hazard?

Debris flood: rapid muddy flow of intermediate consistency between mudflow and stream flood. Highly destructive in urban areas of southern California.

What is an *alpine debris avalanche?* Where does it occur? Of what materials is it composed? How fast does it travel?

Alpine debris avalanche: very fast down-valley flow of mixture of rock waste and snow set loose from high place in alpine mountains.

Landslides

Landslides 292

What is a *landslide?*

Landslide: rapid sliding of large masses of bedrock on steep mountain slopes or from high cliffs.

What are the two basic forms of landslide?

Two basic forms of landslide: (1) *rockslide; (2) slump block.*

What is a *rockslide?*

Rockslide: form of landslide consisting of slippage of a bedrock mass on a sloping fracture plane; resembles motion of a sled. Fracture plane may be bedding plane or fault plane.

What is a *slump block?*

Slump block: block of bedrock that moves down to a lower position of rest on a curved slip plane.

Rockslides

Rockslides 293

Where do rockslides occur?

Large rockslides occur where steep mountain slopes exist, especially on walls of glacial troughs and fiords in great alpine ranges of glove.

Slump Blocks

Slump Blocks 293

Under what geologic conditions do large slump blocks form?

Large slump blocks are typical of cliffs of massive sedimentary strata (sandstone, limestone) or basalt, underlain by weak shale or clay formations.

GEOMORPHIC ASPECTS OF THE ARCTIC TUNDRA

GEOMORPHIC ASPECTS OF THE ARCTIC TUNDRA 295

Arctic Permafrost

Arctic Permafrost 295

What is *permafrost?* Where is permafrost found?

Permafrost: the condition of permanently frozen water within soil and regolith.

What is the *active layer?*

Active layer: shallow surface layer subject to seasonal thawing in permafrost regions.

What three zones of permafrost are recognized? Where are they?

Continuous, discontinuous, and sporadic zones from north to south.

How deep is permafrost in the continuous zone?

Permafrost depth up to 300–450 m (1000–1500 ft) in continuous zone.

Features of the Active Surface Layer

Features of the Active Surface Layer 295

How are coarse rock fragments affected by continued frost action?

Pebbles and cobbles are forced to surface by ground-ice heave, then moved laterally to become sorted into *stone polygons.*

What are *stone polygons?*

Stone polygons: ringlike networks of coarse rock fragments (same as stone rings, stone nets).

What forms are produced by *ice wedges?*

Ice wedges: vertical, wall-like bodies of ground ice, often tapering downward, occupying shrinkage cracks in silt of *permafrost* regions. Ice wedges are arranged in *ice-wedge polygons.*

What are *ice-wedge polygons?*

Ice-wedge polygons: polygonal networks of ice wedges.

What is *patterned ground?*

Patterned ground: general term for a ground surface that bears polygonal or ringlike features, including stone polygons and ice-wedge polygons.

What is a *pingo?*

Pingo: conspicuous conical mound or circular hill having a core of ice, found on plains of the arctic tundra where permafrost is present. Pingos may be as high as 50 m (165 ft).

What is *solifluction?* Where is it a dominant process? What forms does it produce?

Solifluction: tundra variety of earthflow in which saturated thawed layer flows slowly to produce multiple terraces and lobes on hillsides.

Environmental Problems of Permafrost

Environmental Problems of Permafrost 296

What is *thermal erosion?*

Thermal erosion: spontaneous turning of frozen soil to mud and removal by flowing water, as permafrost melts.

What environmental concerns have been expressed over construction of the Trans-Alaska Pipeline? How may the tundra ecosystem sustain damage?

Trans-Alaska Pipeline (TAPS): hazards of melting of permafrost by hot oil in pipe. Spills may extensively damage ecosystem.

SCARIFICATION OF THE LAND

SCARIFICATION OF THE LAND 297

For what two basic purposes do humans act as geomorphic agents.

Humans act as geomorphic agents in two ways: (1) by extracting mineral resources; (2) by reorganizing terrain into new forms to support various industrial and residential structures.

What is *scarification?*

Scarification: general environmental impact term for human-made excavations and other land disturbances produced for purposes of extracting or processing mineral resources.

CHAPTER 15—SAMPLE OBJECTIVE TEST QUESTIONS

A. MATCHING

1.	block separation	a. _______	rock sea
2.	saprolite	b. _______	beach sand
3.	outcrop	c. _______	carbonic acid
4.	transported regolith	d. _______	caverns
5.	sheeting structure	e. _______	joints
6.	solifluction	f. _______	desert gravel
7.	felsenmeer	g. _______	residual regolith
8.	travertine	h. _______	unloading
9.	grus	i. _______	bedrock
10.	chemical weathering	j. _______	tundra

B. MULTIPLE CHOICE

1. Which of the following is not a form of physical rock breakup?

   ________a. karst development.

   ________b. exfoliation.

   ________c. granular disintegration.

   ________d. shattering.

2. Which of the following is not typical of cold arctic and alpine regions?

   ________a. permafrost.

   ________b. saprolite.

   ________c. pingo.

   ________d. solifluction.

3. Sheeting structure

   ________a. results from the process of unloading.

   ________b. leads to spheroidal weathering.

   ________c. is essential for development of saprolite.

   ________d. is found in large earthflows.

C. COMPLETION

1. The dropping of loose rock fragments from a cliff produces a ________________________.
2. A prominent, rounded hill summit of bare rock, having sheeting structure is called an ______ ________________________.

3. A thick layer of residual regolith often present in regions of warm, humid climate, is called___
   ________________________.

4. A variety of earthflow found on tundra slopes is ____.

5. Severe human-made landscape alteration involving removal and deposit of earth materials is called ____________________________________.

CHAPTER 16

RUNOFF, STREAMS, AND GROUND WATER

PAGE

RUNOFF AS A VITAL RESOURCE

RUNOFF AS A VITAL RESOURCE 298

Why is runoff a vital resource?

Runoff is a vital resource because industrialized society requires enormous quantities of fresh water.

What is the role of water pollution?

Water pollution tends to increase as populations grow and urbanization spreads, reducing supplies of pure, fresh water.

FORMS OF OVERLAND FLOW

FORMS OF OVERLAND FLOW 298

Trace the surface paths of flow of surplus water on the lands.

Surplus water on the lands follows a system of flow paths leading to lower levels.

What is *interception?* What may happen to intercepted water?

Interception: the catching and holding of precipitation above the ground surface on leaves and stems of plants. Intercepted water may evaporate directly.

What is *depression storage?*

Depression storage: temporary holding of precipitation in minor surface depressions of the soil surface (same as *surface detention*). Some may evaporate; remainder infiltrates the soil.

What is *overland flow?* What forms does overland flow take?

Overland flow: runoff down land slopes in broadly distributed sheets. Forms of overland flow: thin films (*sheet flow*), rivulets connecting hollows, subdivided flow in grass, flows beneath leaf mat in forests.

What is *stream flow?* What forms does overland flow take?

Stream flow (channel flow): water flow in a narrow channel confined by lateral banks.

How is overland flow measured?

Overland flow is measured in units of depth of water (cm, in.) per hour.

What formula expresses the rate of production of overland flow?

Rate of production of overland flow = rate of precipitation minus rate of infiltration.

INTERFLOW

What is *interflow?* What soil structure is favorable to the occurrence of interflow?

INTERFLOW 299

Interflow: movement of soil water through a permeable soil horizon or other shallow permeable layer in a downslope direction parallel with the ground surface. (Also called *throughflow.)* B horizon of clay accumulation (argillic horizon) may deflect infiltration to form interflow moving through a sandy A horizon.

DRAINAGE SYSTEMS

What is a *stream?* What causes the water to flow? How is stream flow different from overland flow?

What is the concept of organization of runoff within the drainage system with respect to requirements of runoff and load?

Define *drainage system.* What forms the outer limit, or perimeter of the system? Toward what point does it converge?

Define *drainage basin.* What is the shape of the typical drainage basin?

What feature represents the unit cell of the drainage system?

DRAINAGE SYSTEMS 299

Stream: a long, narrow body of flowing water occupying a trenchlike depression, or channel, and moving to lower levels under the force of gravity.

Seeking to escape to lower levels, runoff becomes organized into a drainage system adapted in form to dispose as efficiently as possible of runoff and load of mineral particles.

Drainage system: a branched network of stream channels and the adjacent land slopes, bounded by a drainage divide and converging to a single channel at the outlet.

Drainage basin: entire land surface within a drainage divide contributing to outflow by a single channel. Most natural drainage basins are pear-shaped in outline, pointed at the outlet.

Unit cell of the drainage system is the fingertip upper end of the channel together with the surface area draining into it.

STREAM CHANNEL GEOMETRY

What is a *stream channel?*

In what ways is the geometry of a stream channel described?

What is *stream gradient?*

STREAM CHANNEL GEOMETRY 300

Stream channel: a narrow trough, shaped by forces of flowing stream water. Channel is adjusted to most effective shape for moving water and sediment supplied to stream.

Stream channel geometry: depth (d), width (w), cross-sectional area (A), *wetted perimeter (P).*

Stream gradient: rate of fall of altitude of stream surface in the downstream direction, stated in m/km (ft/mi).

STREAM FLOW

Describe the pattern of water flow in a stream. Where is the friction greatest? Where is flow fastest? where slowest? In what state of motion is the water?

STREAM FLOW 300

Flow pattern within a stream: friction impedes flow close to channel; flow is fastest near center, close to water surface. Flow is usually in turbulent state.

What is *turbulence?* What directions do the water motions take? What is the importance of turbulence in stream flow? In what way does average flow conform with channel configuration?

Turbulence: water motions consisting of innumerable eddies that continually form and dissolve, creating upward, downward, and sidewise flow directions. Turbulence enables fine sediment to be held suspended in the flow. Average flow is parallel with bed and banks, measured over long time period.

Define *mean velocity.*

Mean velocity: a mean, or average value of velocity of flow representative of the entire stream cross section.

STREAM DISCHARGE

STREAM DISCHARGE 301

Define *discharge.* What symbol represents discharge? In what units is discharge stated? Give both metric and English units.

Discharge: (symbol Q) volume of flow moving through a given cross section of a stream in a given unit of time. Units are cubic meters per second (cms) or cubic feet per second (cfs).

How are discharge, area, and mean velocity related? Give the equation for this relationship.

Relations of discharge, area, and mean velocity: Discharge is equal to cross-section area times mean velocity. Equation: $Q = AV$.

How is velocity related to both gradient and cross-sectional area? How does flow velocity and cross section of a steep gradient stream differ from that of a stream of gentle gradient?

Relations of velocity to gradient and cross-sectional area: Flow is faster with steeper gradient, hence cross-sectional area is reduced. Flow is slower with gentler gradient, hence cross-sectional area is increased. Q must equal AV at all times.

STREAM GAUGING

STREAM GAUGING 301

What is *stream gauging?*

Stream gauging: measurement of stream discharge, mean velocity, and depth continuously or at intervals over a long period of time at a selected station.

What instruments and equipment are used in stream gauging?

Stream-gauging requires measurement of *stream stage,* using a *staff gauge,* and of velocity, using a *current meter.*

Define *stream stage.*

Stream stage: height or elevation of the surface of a stream at any given instant.

Describe a *staff gauge.*

Staff gauge: a graduated vertical scale used to measure directly the stream stage.

How can stream stage by measured automatically?

Automatic and continuous measurement of stream stage is made in a stilling tower, using a float-type automatically recording gauge.

What is a *current meter?*

Current meter: device to measure the velocity of flow of water at a given point in a stream. Commonly used meter has propellor blades whose speed of revolution is proportional to current speed.

How is discharge measured?

Discharge (Q) is determined by measuring current speed at sample points on a cross-sectional grid pattern, calculating mean velocity (V), and multiplying by area (A) of cross-section measured from bed profile.

STREAM FLOW AND PRECIPITATION

STREAM FLOW AND PRECIPITATION 302

What is the concept of lag time of runoff with respect to precipitation? How is size of drainage basin involved?

Peak discharge of a stream lags in time behind the period of heavy runoff-producing precipitation; that time lag is greater for larger drainage basins than for small.

What is a *hydrograph?*

Hydrograph: a graph showing how stream discharge (vertical axis) changes with passage of time (horizontal axis). Example from a small water shed shows rapid rise in discharge a few hours after precipitation period, followed by gradual decline over three-day period.

What is the concept behind delay of runoff following precipitation? Where is the flow delayed?

Part of the precipitation from a rainstorm percolates down to the ground water zone, where it is held in storage and given up slowly to streams as *base flow.*

What is *base flow?* Where does it come from? In what type of climate is base flow found?

Base flow: portion of the stream flow supplied by ground water seepage. Found only in streams of moist climates where water table is high.

Define *lag time.* How is it measured? What is responsible for the lag? How does lag time relate to size of watershed?

Lag time: interval of time between occurrence of precipitation and peak discharge. Measured as time elapsed from center of mass of precipitation to center of mass of runoff. Lag time represents delay factor because channels act as temporary reservoirs. Lag time generally increases with increase in area of watershed.

BASE FLOW AND OVERLAND FLOW

BASE FLOW AND OVERLAND FLOW 303

What concept related hydrograph of a large stream to sources of supply?

Hydrographs of large streams in humid climates show effects of two sources of water supply: (1) base flow; (2) overland flow.

What causes sharp peaks in the hydrograph of a large river? How does base flow change from winter to summer?

Example of large river (Chattahoochee River) shows many sharp peaks caused by rainstorms, superimposed on base flow curve. Base flow rises in winter and spring, declines through summer and fall.

What factors of climate influence the hydrograph of a large river such as the Missouri?

Missouri River shows annual cycle with high flow rate in spring, derived from snowmelt in high mountain watersheds. Low flow in winter reflects frozen state of soil water.

FLOODS

What is a *flood?* Where does the water accumulate?

What is a *floodplain?* How often do floods occur for a typical river in a humid climate? Can a floodplain support vegetation? Explain.

Define *flood stage.*

Define *bank-full stage.*

FLOODS 304

Flood: stream flow that cannot be accommodated within the channel and spreads over the banks to inundate the *floodplain.*

Floodplain: belt of low, flat ground, present on one or both sides of a stream channel, subject to inundation on the average of about once per year during season of maximum discharge. Floodplains sustain forests or crop agriculture during season between floods.

Flood stage: a designated stage for a particular point on a river greater than which *overbank flooding* may be expected. Immediately at or below flood stage is the *bank-full stage.*

Bank-full stage: stream stage corresponding with flood stage, at which time discharge is contained entirely within the limits of the stream channel.

DOWNSTREAM PROGRESS OF A FLOOD WAVE

What is a *flood crest?*

What is the *flood wave?*

How is time of occurrence of the flood crest related to downstream distance?

What effect does increasing discharge downstream have on the flood wave and its crest?

DOWNSTREAM PROGRESS OF A FLOOD WAVE 304

Flood crest: maximum stream stage attained during the passage of a flood. Crest moves downstream with the *flood wave.*

Flood wave: time sequence of rising and falling stream stage during the passage of a single flood.

Flood crest occurs at progressively later times at points farther downstream. (Lag time increases downstream.)

At points progressively farther downstream, flood peak discharge increases, while the flood wave becomes more attenuated (drawn out over longer period of time).

FLOOD PREDICTION

What organization is in charge of flood forecasts and warning in the United States?

What is a monthly flood expectancy graph? What information does it give?

FLOOD PREDICTION 305

National Weather Service maintains flood forecast and warning service. Warnings are publicized by all possible means.

Monthly flood expectancy graph: graph showing percentage of years in which flood stage and other stages have occurred, establishing probability of floods occurring in any given month. Example: Mississippi River has topped flood stage only in same six consecutive months during period of record.

FLASH FLOODS AND THEIR PREDICTION

Define *flash floods.*

What three weather situations produce flash floods?

How do flash floods in forested regions differ from those of the arid west?

FLASH FLOODS AND THEIR PREDICTION 306

Flash floods: sudden and short-lived overbank discharges following torrential rainstorm.

Flash floods are produced (1) by hurricanes that penetrate continental interiors; (2) by a slow-moving or stagnant weather front that imports large quantities of moist tropical air; and (3) by orographic precipitation in mountainous regions.

In heavily forested watersheds, the flash flood carries little bedload and is highly erosive. In arid western watersheds, great quantities of rock debris are carried by the floodwaters.

HYDROLOGIC EFFECTS OF URBANIZATION

What is the concept of urbanization as a cause of change in hydrologic conditions, such as floods and lag times?

In what way can flooding be taken into account in urban land-use planning?

HYDROLOGIC EFFECTS OF URBANIZATION 306

Urbanization, by rendering ground surface impermeable, causes greater runoff, with higher flood peaks, and shorter lag times in nearby streams.

Land-use planning should include assessment of vulnerability or urban land to flooding, including effects of expanded urbanization.

GROUND WATER

What is *ground water?* What zone does the ground water occupy?

Define *saturated zone.*

What is the *water table?*

What is the *capillary fringe?*

How can the height and surface distribution of the water table be determined?

GROUND WATER 308

Ground water: that part of the subsurface water which fully saturates pore spaces of the bedrock and regolith. Ground water occupies the *saturated zone.*

Saturated zone: zone in which pores are not fully saturated. Water is held in capillary films adhering to mineral surfaces. *Unsaturated zone* lies above saturated zone.

Water table: upper boundary surface of the saturated zone.

Capillary fringe: thin layer immediately above the water table in which water has been drawn upward from the ground water zone beneath by capillary force.

Mapping of water table is done by plotting height of water standing in wells.

Where is the water table highest? In what direction does it slope to lower levels? What happens at points where the water table intersects the ground surface? Where does the water seep out? How does percolation of water from above affect the water table? What condition results?

Water table is highest beneath hilltops and divides, slopes down to low points at streams, lakes, and marshes. Here water table intersects the surface, allowing water to seep out. Percolation from above tends to raise level of water table; seepage tends to lower it.

GROUND-WATER MOVEMENT

GROUND-WATER MOVEMENT 308

What is a *hydraulic head*? How does the hydraulic head affect flow?

Hydraulic head: difference in level of water table from one point to another; causes flow toward lower level.

How does hydraulic head change when precipitation increases? during drought?

Hydraulic head increases in periods of high precipitation; declines in drought periods.

Describe the seasonal cycle of water-table fluctuations in a humid climate.

Seasonal cycle of fluctuation of water-table in humid midlatitude climate shows rise in late winter and spring due to *ground water recharge,* a decline in summer and fall.

Define *ground-water recharge.*

Ground-water recharge: replenishment of ground water by downward movement of surface water through the unsaturated zone, or from stream channels, or by use of recharge wells.

What function does outward seepage of ground water perform in the hydrologic cycle?

Seepage of ground water feeds streams, ponds, and marshes; completing subsurface phase of hydrologic cycle.

Describe the paths of ground-water flow. Where is motion fastest? where slowest?

Paths of ground-water flow: descend deeply from under divides, then curve back up to emerge in streams. Motion very slow in deep paths. Flow fastest in shallow paths and near streams.

GROUND WATER IN STRATA

GROUND WATER IN STRATA 309

How do sedimentary strata exert a control over ground water?

Sedimentary strata can exert a strong control over the storage and movement of ground water.

What is meant by *porosity* of rock?

Porosity: the ratio of open spaces (cavities, voids) to the total volume of bedrock or regolith. Example: sand or gravel usually has high relative porosity.

What is *permeability?*

Permeability: property of relative ease of movement of ground water through rock or regolith.

What is an *aquifer?* What kinds of rock usually make good aquifers?

Aquifer: a rock layer that has high permeability. Sandstone or sand layers are common types of aquifers; tephra forms excellent aquifers.

What is an *aquiclude?* What kinds of rocks commonly make aquicludes?

Aquiclude: a rock layer that impedes the flow of ground water. Shale and clay beds are common types of aquicludes. Permeability is low.

How is a *perched water table* produced?

Aquicludes can block downward percolation of water, creating a *perched water table.*

What is an *unconfined aquifer?*

Unconfined aquifer: aquifer capable of receiving recharge directly upon its upper surface by vertical recharge.

What is a *spring?* What is the source of water of a spring?

Spring: any outflow of ground water at the ground surface occurring naturally. Springs may occur where an aquifer intersects the side of a hill or mountain.

What is *artesian flow?*

Artesian flow: spontaneous upward movement of ground water in a well or spring, coming from a confined aquifer.

Describe a *confined aquifer.*

Confined aquifer: permeable rock layer capped by an impermeable rock layer, or aquiclude.

How high will artesian water rise in a well?

Rise of aretesian water in a well is limited by the *hydrostatic pressure surface,* which declines in elevation away from the region of surfaced recharge.

What is an *artesian well?*

Artesian well: well in which water rises to the height of the hydrostatic pressure surface, and which may pour out upon the ground surface.

In what way is artesian water an important resource?

Artesian water has been an important water resource in many areas in the past, but it has been heavily overdrafted and pumps must now be used.

GROUND-WATER WITHDRAWAL

GROUND-WATER WITHDRAWAL 311

What concept relates human use of ground water with the workings of the hydrologic cycle?

Human's heavy withdrawals of ground water have greatly altered the natural hydrologic balance.

How are wells put down into the earth? What capacity does a large well have? What is the effect of pumping of water upon the water table?

Wells: may be dug or drilled. Drilled wells, called *tube wells,* use metal casing. Large industrial and irrigation wells can yield millions of liters per day. Pumping lowers water table.

What is a *cone of depression?* What causes a cone of depression?

Cone of depression: conical configuration of water table around a pumped well, representing the extent of lowering by withdrawal.

What is *drawdown?* What effect has drawdown upon rate of flow of water to the well? Explain.

Drawdown: difference in height between cone base and original water table surface. Drawdown steepens the hydraulic gradient, increases speed of flow toward well.

What are the effects of sustained heavy pumping of ground water?

Effects of sustained heavy pumping: where wells are close together, entire water table may be lowered many meters. Depletion of ground water may greatly exceed rate of recharge, exhausting the water resource.

How is ground water naturally recharged in dry climates?

In dry climates, streams flowing across alluvium lose water by seepage to ground water body beneath; these are *influent streams.* In moist climates *effluent streams* receive ground water that seeps into the channel.

How can ground water be artificially recharged?

Artificial recharge of ground water can be made by use of *recharge wells,* into which surplus surface is pumped.

ENVIRONMENTAL PROBLEMS RELATED TO GROUND-WATER WITHDRAWAL 312

POLLUTION OF GROUND WATER

POLLUTION OF GROUND WATER 312

What concept relates ground water pollution to solid waste disposal?

Widespread practice of solid waste disposal into ground leads to pollution of ground water and contamination of water supplies.

What is *sanitary landfill?*

Sanitary landfill: method of burial of waste, without burning, under a cover of packed earth (clay, sand).

What is *leachate?* How does it travel? What environmental damage can leachate do?

Leachate: solution of ions derived from buried waste by percolating precipitation, moving downward to reach water table. Leachate may travel with ground water to reach wells or streams.

How can a supply well be contaminated by leachate?

Supply well can be contaminated by leachate, which forms a ground water mound, setting up a hydraulic gradient toward the cone of depression of the well.

How can the movement of leachate be detected?

Movement of leachate can be detected by use of monitor wells.

SALTWATER INTRUSION

SALTWATER INTRUSION 313

What is saltwater intrusion? Where and how does it occur?

Saltwater intrusion is the landward movement of salt ground water in the coastal zone, replacing fresh ground water and causing contamination of supply wells.

What configuration does the body of fresh ground water have beneath an island or peninsula?

Beneath an island or peninsula, the fresh ground water body has form of a doubly-convex lens, floating on salt ground water.

What is the relationship of height of water to depth of base of the fresh ground water lens?

Depth to base of fresh water lens is about 40 times the height of the water table, expressing the 40 to 41 ratio of density of fresh and salt water.

How does saltwater contamination occur?

Withdrawal of fresh water causes upward and landward migration of salt water-fresh water interface.

LAND SUBSIDENCE FROM GROUND-WATER WITHDRAWAL

What is land subsidence? Why does it result from ground-water withdrawal?

Name some examples of cities where land subsidence from ground-water withdrawal has created environmental problems.

LAND SUBSIDENCE FROM GROUND-WATER WITHDRAWAL 314

Land subsidence occurs when the land surface over broad area sinks. It results from ground-water withdrawal when sediments contract in volume as water is drained out.

Houston, Texas; Mexico City, Venice, Italy; Bangkok, Thailand.

LAKES AND PONDS

What is a *lake?*

By what processes may a lake disappear?

How is the level of a lake related to the water table?

What is the origin of freshwater marshes and swamps?

LAKES AND PONDS 315

Lake: body of standing water within a land area. It has an exposed free upper water surface and may be fresh or saline, but without an influx of seawater. Term can include pond, swamp, and marsh.

Lakes may disappear by lowering of the outlet channel, by accumulation of sediment and organic matter, or by excessive evaporation.

In moist climates, water level of lakes and ponds coincides with water table in surrounding upland area.

Freshwater marshes and swamps occur where the ground water table is exposed at the surface in broad topographic depressions.

WATER BALANCE OF LAKES AND RESERVOIRS

What three sources of input provide water to a lake or reservoir?

What three sources of output reduce the volume of water in a lake or reservoir?

What is an *evaporating pan?*

WATER BALANCE OF LAKES AND RESERVOIRS 316

Input to a lake or reservoir is provided by runoff from streams; precipitation; and ground-water subsurface inflow.

Output that reduces the volume of water in a lake or reservoir are runoff to an existing stream or streams; evaporation; and ground-water subsurface outflow.

Evaporating pan: a large circular container from which water is allowed to evaporate. It is used to estimate the amount of evaporation from the free water surface of a lake or reservoir.

SALINE LAKES AND SALT FLATS

How does a saline inland lake respond to climate changes?

SALINE LAKES AND SALT FLATS 317

Saline inland lakes expand when influx of stream flow increases. Increased evaporation rate from the expanded surface will balance the increased rate of input.

What is an *exotic river?*

Exotic river: stream sustained in its flow across an arid region by means of runoff derived from distant upland where a water surplus exists. Examples: Colorado River, Nile River.

How does evaporation in an arid climate affect human-made reservoirs?

In arid regions much of the water held in reservoirs behind dams is lost by direct evaporation.

THE ARAL SEA—A DYING SALINE LAKE 317

Where is the Aral Sea? How large is it? Does it have an outlet?

The Aral Sea lies east of the Caspian Sea in the Soviet Union. It was the world's fourth largest lake. It has no outlet.

Why did the Aral Sea shrink? By how much did its volume shrink?

The Aral Sea shrank because most of the water flowing into it from two large rivers was diverted into a canal system in order to provide water for cotton crops. The volume was reduced by two-thirds.

What was the effect on salinity? On the lake's ecosystem?

The salinity increased from 10% to 27%. Most native fish species disappeared. Delta islands were degraded. Many species of animals disappeared.

DESERT IRRIGATION AND SALINIZATION

DESERT IRRIGATION AND SALINIZATION 318

How are irrigation systems set up? What is the water source?

Irrigation systems divert water of exotic rivers.

Define *salinization.* How does salinization occur? In what way is it harmful to crops?

Salinization: buildup of salts in soil through persistent evaporation or irrigation water or shallow ground water. May ruin soil for crop production.

Define *waterlogging.* What causes waterlogging?

Waterlogging: rise of shallow water table to reach root zone of crops and ruin agriculture.

Describe the harmful effect of salinization and waterlogging upon irrigation areas. Give an example.

Salinization and waterlogging threaten major irrigation systems in deserts. Example: Indus R. system in Pakistan. Great caution advised in future expansion of desert irrigation schemes.

CHEMICAL SOURCES OF WATER POLLUTION 319

Name the important chemical pollutants. Six ions should be listed. What is the source of nitrate ion? Where does the phosphate ion come from? What undesirable effect does it produce? What is the source of chloride? of sulfate?

Common chemical pollutants: ions of sulfate, nitrate, phosphate, chloride, sodium, and calcium. Nitrate is highly toxic, comes from sewage and fertilizers. Phosphate from detergents is nutrient, causing *eutrophication* in streams and lakes. Chloride from deicing salt. Sulfate from atmospheric fallout.

Define *eutrophication*. What causes eutrophication?

Eutrophication: excessive growth of algae and other primary producers in streams and lakes as a result of inputs of nutrient ions, especially phosphate and nitrate.

What is *acid mine drainage?* What acid is present? Where does it come from? What harmful effects result from acid mine drainage?

Acid mine drainage: sulfuric acid effluent from coal mines and tailings or spoil ridges from strip mining. Acid is derived from sulfur-rich minerals in coal seams. Acid causes corrosion damage, death of fish in downstream locations.

What is the source of toxic metals as water pollutants? Name five toxic metals.

Toxic metals as a form of water pollution occurs from industrial wastes. Dangerous ions include zinc, lead, arsenic, copper, and aluminum.

What is *thermal pollution?* What is the source of thermal pollution? In what way is it harmful to the environment?

Thermal pollution: heated water discharge from power plants entering a stream or lake (or saltwater estuary). May have serious impact upon organisms in water.

THE RUNOFF ENERGY SYSTEM

THE RUNOFF ENERGY SYSTEM 320

What is the runoff energy system? How is it powered?

The runoff energy system is an open energy system in which work is done as water moves to lower levels and encounters frictional resistance. It is powered by gravity.

What is the energy input to the runoff energy system?

The energy input to the runoff energy system is the potential energy present in precipitation that arrives at the land surface.

What are the two subsystems of the runoff energy system.

The two subsystems are the surface water subsystem and the ground-water subsystem.

What is the role of kinetic energy in the surface water subsystem?

Kinetic energy is produced from potential energy as water moves downhill under the influence of gravity. The kinetic energy is transformed to heat in friction against the riverbed below and air above. This heat is lost by radiation, conduction, or evaporation.

What is the role of kinetic energy in the ground-water subsystem?

The ground-water subsystem receives potential energy during recharge, and this is changed to kinetic energy as ground water flows in response to hydraulic gradients.

CHAPTER 16—SAMPLE OBJECTIVE TEST QUESTIONS

A. MATCHING

1.	ground water	a. ________	channel
2.	exotic river	b. ________	stage
3.	hydraulic head	c. ________	leachate
4.	drawdown	d. ________	desert
5.	sanitary landfill	e. ________	saturated zone
6.	lake	f. ________	gradient
7.	stream flow	g. ________	pumping
8.	stream velocity	h. ________	water table
9.	stream gauging	i. ________	phosphate ion
10.	eutrophication	j. ________	pond

B. MULTIPLE CHOICE

1. Water in the unsaturated zone

________a. is unavailable to plants.

________b. completely fills all pore spaces.

________c. is held in capillary films.

________d. is derived from the saturated zone below.

2. Stream discharge

________a. is greater on a steeper gradient.

________b. is equal to the product of area and mean velocity.

________c. is increased by turbulence within the stream.

________d. is equal to the mean velocity times the gradient.

3. The lag time between precipitation period and peak discharge

________a. is increased by urbanization.

________b. is decreased by urbanization.

________c. is shorter for a large drainage basin than for a small one.

________d. depends upon the hydraulic head.

C. COMPLETION

1. A belt of low, flat ground bordering a river channel and subject to inundation is called a ________________________.

2. Excessive growth of algae in a stream or lake because of an input of nutrient ions is called ________________________.

3. The acid found in acid mine drainage is ________________________ acid.

4. Deicing salts are a source of the ________________________ ion as a pollutant in surface water.

5. A graph showing how stream discharge varies with time is called a ________________________ ________________________.

6. Stream flow provided by ground water seepage is called ________________________ ________________________.

CHAPTER 17

LANDFORMS MADE BY RUNNING WATER

PAGE

What are the *fluid agents?* What work does each perform? Are liquids and gases both classed as fluids?

Fluid agents: natural agents which erode, transport, and deposit mineral matter and organic matter; namely, running water, waves and currents, glacial ice, and wind.

Define denudation. Toward what goal does denudation work? What is the lower limit to which denudation can be carried out? What becomes of the sediment produced during denudation?

Denudation*:* the total action of all processes by which the exposed rocks of the continents are worn down and the resulting sediments are transported to the sea by the fluid agents. Denudation is an overall lowering of the land surface, tending to reduce the continents to sea-level surfaces.

FLUVIAL PROCESSES AND LANDFORMS

FLUVIAL PROCESSES AND LANDFORMS **322**

What are *fluvial landforms?*

Fluvial landforms: landforms shaped by running water, i.e., by *fluvial processes.*

Define *fluvial processes.* Where are fluvial processes dominant? What functions do fluvial processes perform?

Fluvial processes: running water as overland flow and stream flow, aided by weathering and mass wasting. Fluvial processes and landforms are dominant on the lands. Fluvial processes perform erosion, transportation, and deposition.

What are the two groups of fluvial landforms?

Two groups of fluvial landforms: (1) *erosional landforms*; (2) *depositional landforms.*

What are *erosional landforms?* Give two examples.

Erosional landforms: sequential landforms shaped by progressive removal of the bedrock mass. Examples: valley, ridge.

What are *depositional landforms?* Give two examples.

Depositional landforms: sequential landforms built of rock and soil fragments transported by fluid agents and deposited in another location than the source. Examples: floodplain, fan.

NORMAL AND ACCELERATED SLOPE EROSION

NORMAL AND ACCELERATED SLOPE EROSION 323

What concept deals with erosion as a natural process?

Slow, continuous erosion of the land surface is a natural process, both inevitable and universal.

What is the *geologic norm?* Relate the geologic norm to soil horizons and plant communities.

Geologic norm: stable natural condition in a humid climate in which slow erosion of soil is paced by maintenance of soil horizons bearing a plant community in an equilibrium state.

What is *accelerated erosion?* What is the end result of accelerated erosion?

Accelerated erosion: soil erosion occurring at a rate much faster than soil horizons can be formed from parent matter. Accelerated erosion results in thinning of the soil by removal of the upper horizons and leads to total disappearance of soil.

What causes accelerated erosion? How are humans involved? What is the effect of destruction of vegetation?

Cause of accelerated erosion: human activities, or rare natural events forcing change in plant cover and in physical state of ground surface. Destruction of vegetation allows rain to beat directly on soil, sealing pores and increasing rate of runoff by overland flow.

Describe *splash erosion.* What is the effect of splash erosion?

Splash erosion: lifting of soil particles by impact of raindrops on wet soil surface. Splash erosion seals soil openings, shifts soil slowly downhill.

What concept relates plant cover to overland flow and soil erosion?

Plant cover offers resistance to overland flow, protects soil from removal.

CHANGES IN INFILTRATION CAPACITY

CHANGES IN INFILTRATION CAPACITY 324

What is *infiltration capacity?*

Infiltration capacity: limiting rate at which falling rain or melting snow can be absorbed by a soil surface in the process of infiltration.

How does infiltration capacity change as heavy precipitation continues to fall?

Infiltration capacity is initially high at end of long dry period; falls rapidly after precipitation begins; levels off at nearly constant rate after 1 to 3 hours.

Why does infiltration capacity decline with time?

Infiltration capacity over prolonged periods is high for coarse-textured soils (sands); low for clay-rich soils.

How does land use affect infiltration capacity?

Infiltration capacity can be greatly reduced by poor land-use practices, including overgrazing and deforestation.

LAND USE AND SEDIMENT YIELD

LAND USE AND SEDIMENT YIELD 324

What is *sediment yield?* In what units is it measured?

Sediment yield: quantity of sediment removed by overland flow from a land surface of given unit area in a unit of time. Units are metric tons/hectare.

How does land use influence sediment yield?

In moist midlatitude climate, sediment yield is extremely high on bare, cultivated soil surfaces, moderately high on pastures, low on abandoned fields and depleted forest, very low on fully forested land.

What natural environments are particularly vulnerable to accelerated erosion?

Semiarid climate zones (steppes) are particularly vulnerable to change from normal to accelerated erosion, because plant cover is sparse and easily damaged by overgrazing or fire.

What are *badlands?* What rock type produces badlands? In what climate are badlands self-sustaining?

Badlands: rugged surfaces resembling miniature mountains, developed on weak clay formations by erosion too rapid to permit plant growth. Badlands are self-sustaining in arid climates.

FORMS OF ACCELERATED SOIL EROSION

FORMS OF ACCELERATED SOIL EROSION 325

What concept concerns the relationship of fossil fuels to accelerated soil erosion?

Human uses energy of fossil fuels through machines to maintain bare soils against natural restorative forces of plant growth.

What are the successive stages of surface destruction through accelerated soil erosion?

On cultivated surface destruction of soil progresses from sheet erosion, through rill erosion, to gully formation.

Describe *sheet erosion.* Where does the eroded material go?

Sheet erosion: progressive removal by overland flow of thin, uniform layers of soil. Soil thus removed accumulates as colluvium or travels to streams.

What is *colluvium?*

Colluvium: layer of soil particles constructed at base of hillslope by sheet erosion.

What is *alluvium?*

Alluvium: sediment layer deposited by a stream in valley floor.

What is *rill erosion?*

Rill erosion: formation of numerous closely spaced channels carved in soil of steep hillslopes by runoff from heavy rains.

What are *gullies?*

Gullies: deep, steep-walled trenches eroded in soil and regolith by overland flow concentrated into channel flow in a region experiencing accelerated soil erosion.

GEOLOGIC WORK OF STREAMS

GEOLOGIC WORK OF STREAMS 326

Name the three forms of geologic work of streams.

Three forms of geologic work of streams: *erosion, transportation, deposition*. All three forms are part of a single process.

Define *erosion.*

Erosion: progressive removal of mineral matter from floor and sides of a stream channel.

Define *transportation.*

Transportation: movement of eroded particles by dragging along the stream bed, by suspension in the stream water, or in solution.

Define *deposition.*

Deposition: accumulation of transported particles upon the stream bed, upon the adjacent floodplain, or in a body of standing water.

STREAM EROSION

STREAM EROSION 327

Name the three forms of stream erosion.

Stream erosion takes three forms: *hydraulic action, abrasion*, and *corrosion.*

Define *hydraulic action.* In what materials is it effective? How is the stream affected by hydraulic action?

Hydraulic action: erosion by flowing water exerting impact force and dragging action on bed and banks of channel. Effective in unconsolidated materials, such as alluvium. Hydraulic action causes *bank caving* and rapidly adds sediment to streams in flood.

Define *abrasion.*

Abrasion: erosion of bedrock channel walls by impact of particles carried by stream and by rolling of cobbles over stream bed.

Define *corrosion.*

Corrosion: erosion of bedrock by chemical processes of rock weathering, principally acid reactions upon mineral surfaces.

What is a *pothole?* How is it formed?

Pothole: cylindrical cavity in hard bedrock of a steam channel produced by abrasion of a large spherical or discus-shaped rock fragment rotation within the cavity.

STREAM TRANSPORTATION

STREAM TRANSPORTATION 327

In what three forms does a stream carry its load?

Stream carries its load in three forms: (1) dissolved as chemical ions in solution; (2) in suspension as fine particles; (3) as bedload of coarse particles.

Define *stream load.*

Stream load: the solid matter carried by a stream in solution, suspension, and as bed load.

What is *turbulent suspension?* How does matter in suspension affect the appearance of the stream water?

Turbulent suspension: carrying of clay and silt particles in body of stream, upheld by upward elements of flow in turbulent eddies. Suspension causes stream water to be turbid.

Define *suspended load.*

Suspended load: that part of the stream load carried in suspension.

Define *bed load.* What particle sizes move as bed load?

Bed load: that part of the stream load moving close to the bed by rolling, sliding, and low leaps. Sand, gravel, and cobbles move as bed load.

CHANNEL CHANGES IN FLOOD

Describe the usual cycle of changes in channel cross section as a flood wave passes.

What is the meaning of the scour and upbuilding of a stream channel with changes in discharge?

Define *stream capacity*. Is dissolved matter included in this definition?

Can capacity be satisfied when the channel consists of bedrock? Explain.

How is capacity reached for a channel in thick alluvium. How does capacity change as stage increases?

What concept relates capacity increase to mean velocity increase? Be specific. When do most changes in channel form occur?

How is capacity for suspended load affected by change in discharge?

What happens to stream load as discharge and velocity decline?

CHANNEL CHANGES IN FLOOD 328

As flood wave passes, stream channel in thick alluvium undergoes cyclic changes in cross section. Initial change may be upbuilding of bed as discharge rises. Next, bottom scour and deepening occurs as flood crest is reached. With declining discharge, deposition of alluvium raises and restores bed elevation.

Deepening and upbuilding represent changes in *stream capacity*, which varies with stage and discharge.

Stream capacity: the maximum load of solid matter that can be carried by a steam for a given discharge. Both suspended load and bed load are included in total load. (Dissolved matter is not included.)

Stream with bedrock channel cannot pick up enough alluvial material to satisfy its capacity for bed load.

Stream having channel in thick alluvium can easily pick up enough load from bed and banks to satisfy its capacity as flow increases to flood stage.

Concept: capacity for bed load increases rapidly (about as third to fourth power) with increase in mean velocity; therefore most changes in form of channel occur in flood stage.

Suspended load increases enormously as discharge increases, because of increased stream turbulence. (Tenfold increase in discharge brings a 100-fold increase in suspended load.)

Declining discharge and velocity cause cessation of motion of bed load, forming channel bars. Clay of colloidal sizes continues in suspension to ocean or lake.

SUSPENDED SEDIMENT LOAD OF LARGE RIVERS

What factors influence the suspended sediment load of large rivers?

What is noteworthy about the suspended load of the Yellow River?

Why are sediment yields of the Amazon and Congo rivers comparatively small?

SUSPENDED SEDIMENT LOAD OF LARGE RIVERS 329

Great rivers show an enormous range in quantity of suspended load, depending upon climate and land-surface properties.

Yellow River, China, has highest sediment load and sediment yield, because of semiarid climate and exposure of highly erodible wind-deposited silt (loess).

Sediment load can fill large reservoirs behind dams, ending useful life of structure. Trapping of sediment can induce downstream bed scour and channel changes.

CONCEPT OF EQUILIBRIUM AND THE GRADED STREAM 330

What concept relates stream gradient to the necessity for transport of load? What time spans are involved?

Over long spans of time, a stream adjusts the gradient of its channel to provide for transport of all solid load supplied to it by tributaries and watershed surfaces.

What is a *graded stream?* Over what time spans does this condition apply? When does it not apply?

Graded stream: stream with gradient adjusted so as to achieve balanced state in which average bed load transport is matched to average bed load input. Condition of grade is an average state over many years of time; does not apply to rapidly fluctuating activities as river stage rises and falls.

Describe the typical form of the longitudinal profile of a graded stream.

Graded stream shows a *longitudinal profile* that is smoothly up-concave in form. Gradient decreases downstream because stream of larger discharge is more efficient in transport of its given bed load.

Define *longitudinal profile.*

Longitudinal profile: graphic representation of the descending course of a stream or stream channel from upper end to lower end; altitude is plotted on vertical scale, downstream distance on horizontal scale.

EVOLUTION OF A GRADED STREAM 331

What initial conditions can be visualized in studying the evolution of a graded stream?

Initial conditions of stream development consist of ungraded channel with irregularities resulting from geologic history of recent crustal uplift with faulting (crustal fracturing).

List the steps in adjustment of a graded stream. What landforms are produced?

Steps in adjustment of ungraded stream: falls and rapids quickly eliminated, canyon or gorge formed.

What is a *waterfall.*

Waterfall: abrupt descent of stream flow over a bedrock step.

What are *rapids?*

Rapids: fast descent of stream flow over steeply inclined portion of bedrock channel.

Define *gorge, canyon.*

Gorge, or *canyon:* steep-sided bedrock valley with narrow floor limited to width of stream channel.

What concept relates change in input of load to change in stream capacity? What does this have to do with the graded condition?

As stream develops, input of load in increasing while capacity of stream to carry load is decreasing, so that a time comes when load matches capacity and the stream is graded.

Define *graded profile.*

Graded profile: the smoothly descending profile of a stream in the graded condition.

What concept concerns the maintenance of a constant level by a graded stream? What change of channel can occur freely?

The graded stream on the average maintains a constant level or altitude at any given point on its course, but at the same time can shift sidewise.

Define *lateral cutting.* What landform is produced by lateral cutting? In what form does this feature first appear. What features develop later?

Lateral cutting: sidewise shifting of the stream channel by undercutting of bank on the outsides of bends. Lateral cutting leads to formation of a floodplain, which first appears as crescentic areas of low ground on insides of bends. Meanders develop later.

Define *alluvial meanders.*

Alluvial meanders: sinuous bends of a graded stream flowing in the alluvial deposit of a floodplain.

What further changes take place in the graded stream profile as time passes?

With passage of long spans of time, the graded profile is gradually lowered in elevation and gradient diminishes. Profile gradually approaches *base level.*

What is *base level?*

Base level: lower limiting surface or level that can ultimately be attained by a stream under conditions of stability of the earth's crust and sea level; an imagined surface equivalent to sea level projected inland.

Why is base level never reached by a stream?

Because further reduction in elevation of stream becomes slower and slower, base level cannot be attained. Usually, crustal uplift upsets the graded condition.

ENVIRONMENTAL SIGNIFICANCE OF GORGES AND WATERFALLS

ENVIRONMENTAL SIGNIFICANCE OF GORGES AND WATERFALLS 333

How do gorges influence the placement of transportation routes?

Use of gorges for roads and railroads difficult because of lack of space beside stream channel; hazards of floods and rock fall.

In what way are gorges barriers to transportation?

Gorges form barriers to cross-transportation; require expensive bridges.

How are waterfalls used in the generating of electric power?

Large waterfalls, such as Niagara Falls, allow development of large hydroelectric generating plants. Dams usually required to provide necessary drop, as falls are rare on large rivers.

In what way is dam construction destructive of scenic and recreational values of gorges? How are ecosystems affected?

Scenic and recreational value of gorges, falls, rapids as a natural resource: dam construction destroys these values and river ecosystems.

AGGRADATION AND ALLUVIAL TERRACES

AGGRADATION AND ALLUVIAL TERRACES 334

What concept relates the response of graded streams to changes in input of discharge and load? How do channels change to meet such changes?

Graded streams are highly sensitive to changes in input of discharge and load from upstream sources. Channels are readjusted by aggradation or degradation to meet such changes.

What is *aggradation?* How does aggradation alter the channel gradient? How is the stream's capacity changed? What form of channel is produced by aggradation?

Aggradation: a raising of stream channel elevation by deposition of bed load. Aggradation steepens the profile gradient and increases the average capacity of the stream. Aggradation produces a broad, shallow channel of braided form.

Describe a *braided channel.* What grade of bedload is found in a braided channel? How do braided channels change position?

Braided channel: a broad, complex stream channel consisting of multiple threads of fast flow subdividing and rejoining in a manner suggestive of a braided cord. Braided channels have coarse bed materials and shift readily over adjacent low ground.

What causes aggradation? What natural cause has been important in North America?

Causes of aggradation are both natural and human-made. Natural causes include glaciation, which gives large inputs of coarse rock debris from wasting ice margins. Aggradation of this cause was widespread in North America south of glacial limit and caused valleys to be filled with alluvium.

What concept relates channel degradation to change in bed load?

Degradation of a stream channel is a consequence of the reduction in input of bed load materials from upstream sources.

What is *degradation?* In what channel change does it result?

Degradation: lowering of stream channel elevation by progressive removal of bed materials, resulting in trenching of the channel.

What are the causes of degradation?

Causes of degradation: reduction of bed load by growth of plant cover over previously barren watershed surfaces. Reforestation following glaciation was a major event of this type.

Describe *alluvial terraces.* How are alluvial terraces formed?

Alluvial terraces: steplike treads rising from a valley bottom and formed by stream degradation during opening out of an alluvium-filled valley. Meander bend enlargement carves terraces.

What are *rock-defended terraces?*

As older terraces are being removed, some are protected from further attack by outcrops of bedrock and are described as *rock-defended terraces.*

What is the environmental importance of alluvial terraces?

Environmental importance of alluvial terraces: ideal for use as farms, towns, highways, because of flatness and safety from river floods.

AGGRADATION INDUCED BY HUMAN AVTIVITY

AGGRADATION INDUCED BY HUMAN ACTIVITY 336

What is the concept of human involvement in stream aggradation? What activities cause aggradation?

Human activities of cultivation, lumbering, setting fires, strip mining, urban construction, and highway construction denudes land surface, supplying sediment for stream aggradation.

What are the harmful effects of aggradation?

Harmful effects of aggradation: destruction of valley farmlands, filling of reservoirs, burial of ecosystems, silting of tidal estuaries.

LANDFORMS OF FLOODPLAINS

LANDFORMS OF FLOODPLAINS 336

What is an *alluvial river?* How often do floods occur on an alluvial river?

Alluvial river: a river of low gradient flowing upon thick deposits of alluvium and experiencing annual overbank flooding of the adjacent floodplain.

What are *bluffs?*

Bluffs: steeply rising ground surfaces marking the outer limits of the floodplain.

Describe meander development. What forms are produced?

Meander development on a floodplain consists of growth of meander bends followed by cutoffs to produce oxbow lakes.

What is a *cutoff?*

Cutoff: cutting through of the narrow neck of a meander to shorten the course of the river.

What is an *oxbow lake?*

Oxbow lake: a crescent-shaped lake left as a result of a meander cutoff and silting of the ends of the abandoned channel loop.

Describe the channel form of a meander bend. Where is the water deepest?

Channel of a meander is deep on outside of bend, close to bank, where caving occurs. Between bends is shallow crossing with many shifting bars.

What is *downvalley sweep?*

Downvalley sweep: slow, downvalley migration of meander bends under the influence of the slight valley floor gradient.

What is *bar-and-swale?*

Bar-and-swale: assemblage of floodplain landforms consisting of alternate low ridges (bars) and shallow troughs (swales) formed as bar deposits on the inside of a growing meander.

Describe a *natural levee.* When does levee-building take place? Of what sizes of particles is a levee composed?

Natural levee: a belt of higher ground paralleling a meandering stream on both sides of its channel. Levee is built up in times of overbank flooding by deposition of fine sand and silt settling out from flood water.

What is a *backswamp.*

Backswamp: area of low, swampy floodplain lying between the natural levees and the bluffs.

What is a *yazoo stream?*

Yazoo stream: stream that enters upon the floodplain of a larger alluvial river and is forced by the presence of the natural levees to flow far downvalley before making a junction with the larger stream.

What concept relates periodic overbank flooding to soil fertility?

Periodic overbank flooding brings an infusion of dissolved nutrients to the floodplain, sustaining the soil fertility.

ALLUVIAL RIVERS AND HUMAN OCCUPATION

ALLUVIAL RIVERS AND HUMAN OCCUPATION 339

What role have alluvial rivers and their floodplains played in development of the earliest civilizations?

Alluvial river floodplains as sites of earliest civilizations: Nile, Tigris-Euphrates, Indus, Yellow rivers. Dense populations exist today in alluvial valleys of Southeast Asia.

What is the *riverine environment?*

Riverine environment: an alluvial river and its floodplain as an environmental zone of human occupation.

Besides agriculture, what advantages has the riverine environment to human occupation?

Alluvial rivers as natural arteries of transportation have attracted towns and cities, often situated at outsides of meander bends and at points where meanders cut into floodplain bluffs.

FLOOD ABATEMENT MEASURES 340

What two forms of river regulation have been used for flood abatement purposes?

Two basic forms of regulation of river floods: (1) detain and delay runoff on water sheds and in small tributaries; (2) modify lower reaches of river to prevent floodplain inundation.

What forms of watershed treatment are used to achieve flood abatement?

Watershed treatment includes reforestation, erosion control, and small flood-storage dams.

What two basic approaches have been used in modification of the lower river course to achieve reduced flood loss?

Two approaches to modification of lower river: (1) build artificial levee system and create diversion basins; (2) make artificial meander cutoffs to shorten channel and speed up river flow.

What is an *artificial levee (dike)?* What purpose do artificial levees serve?

Artificial levee (dike): earth embankment built parallel with the river channel, usually on crest of natural levee, to contain river in flood stage.

What are the effects of long-continued use and upbuilding of artificial levees?

Long-continued use and upbuilding of artificial levees can lead to upbuilding of river channel, greatly increasing the potential for disastrous flooding when the levees are breached.

How successful has the program of artificial cutoffs of the Mississippi River proved to be?

Shortening of Mississippi River by artificial cutoff of meander loops was initially successful in reducing flood crests, but meander bends have reformed since.

How are floodplain basins used in flood control?

Portions of floodplain enclosed by levees are set aside to receive flood water, thus relieving flood hazard elsewhere.

What alternative exists to the continuation of expensive flood-control measures?

Alternative to continued expensive flood-control structures is adoption of policy of avoidance of human occupation of high-hazard zones.

STREAM REJUVENATION AND ENTRENCHED MEANDERS

STREAM REJUVENATION AND ENTRENCHED MEANDERS 341

What may cause the disruption of a graded river and its profile?

Major disruption of activity of a graded stream can occur through uplift of crust with respect to sea level, causing *rejuvenation.*

Define *rejuvenation.*

Rejuvenation: onset of an episode of degradation by a stream in an attempt to reestablish graded profile at a lower level.

Describe the process of stream rejuvenation.

Rejuvenation begins with development of falls and rapids in lower course, progressing upstream and leading to deep trenching and a steep-walled rock gorge bounded by a *rock terrace.*

What is a *rock terrace?*

Rock terrace: bedrock terrace remaining after the degradation of a stream channel induced by stream rejuvenation.

What are *entrenched meanders?*

Entrenched meanders: winding, sinuous valley bends produced by degradation of a meandering stream with trenching of the bedrock. Entrenched meanders may cut through at narrow neck to leave a *natural bridge* of rock spanning the stream channel.

CHAPTER 17—SAMPLE OBJECTIVE TEST QUESTIONS

A. MATCHING

1.	splash erosion	a. _______	degradation
2.	depositional landform	b. _______	oxbow
3.	badlands	c. _______	bank caving
4.	hydraulic action	d. _______	raindrops
5.	meander	e. _______	braided channel
6.	lake	f. _______	natural levee
7.	aggradation	g. _______	clay formation
8.	alluvial terraces	h. _______	cutoff

B. MULTIPLE CHOICE

1. Capacity of a stream for bed load

________a. depends upon the amount of clay in suspension.

________b. increases as stream velocity increases.

________c. is easily satisfied in a bedrock channel.

________d. is greatest in dry periods between floods.

2. A stream experiencing aggradation is

________a. spreading bed load over the channel floor.

________b. passing over a series of rapids.

________c. typical of streams in a region of humid climate.

________d. undergoing degradation.

3. A natural levee

________a. slopes toward the river channel

________b. is built for the purpose of containing floods.

________c. forces the river out of its banks.

________d. is built during periods of overbank flooding.

C. COMPLETION

1. Limiting rate at which precipitation can enter the soil is the ________________________ capacity.

2. Quantity of sediment removed by overland flow from a unit area of ground in a unit of time is the ________________________ .

3. Stream erosion includes the processes of ________________________, ________________________ action, ________________________, and ________________________ .

4. The maximum load of debris that a stream can carry is the stream ______________________ ______________________.

5. The lower limiting level of stream activity is ______________________ ______________________.

CHAPTER 18

DENUDATION AND CLIMATE

PAGE

Review the early development of the concept of *cycle of denudation.*

Concept of *cycle of denudation,* introduced in late 1800s by W. M. Davis, a geographer, envisioned a succession of developmental stages of an uplifted landmass, starting with an initial stage and progressing through youth, maturity, and reaching old age with the production of a *peneplain.*

Define *peneplain.*

Peneplain: land surface of low elevation and slight relief produced in the late stages of the cycle of denudation. Word derives from "penultimate" and "plain".

CONCEPT OF THE AVAILABLE LANDMASS

CONCEPT OF THE AVAILABLE LANDMASS 343

What is the *available landmass?*

Available landmass: landmass (total bedrock mass) above base level that can be consumed by fluvial denudation.

How does isostatic uplift influence the total available landmass?

Through principle of *isostasy,* available landmass is increased by *isostatic compensation,* to an amount about 5 times the landmass available at the outset of denudation.

What is *isostasy?*

Isostasy: an equilibrium state, resembling flotation, in which crustal masses stand at levels determined by their thickness and density. Equilibrium is achieved by flowage of denser underlying mantle rock.

What is *isostatic compensation?*

Isostatic compensation: slow rising motion of the earth's crust in response to the removal of rock during denudation, as required by isostasy.

TECTONIC UPLIFT AND PLATE TECTONICS

TECTONIC UPLIFT AND PLATE TECTONICS 344

What is tectonic uplift?

Tectonic uplift: Rapid upward movement of crust produced by tectonic activity related to lithospheric plate interaction.

RATES OF TECTONIC UPLIFT

RATES OF TECTONIC UPLIFT 344

How rapid is uplift by tectonic processes?

Tectonic uplift is usually rapid compared with the rate of fluvial denudation. A mountain range might be elevated to a height of 6 km in about one million years.

RATES OF DENUDATION

RATES IF DENUDATION 344

What factors cause place-to-place differences in rates of denudation?

Rate of denudation, measured as depth or thickness of soil or rock removed per unit of time, varies enormously from place to place depending on climate, relief, steepness of slope, and rock type.

On what data are denudation rates based?

Denudation rates are estimated from sediment load of streams and from rates of reservoir sediment accumulation.

What denudation rates can be used in a model of the denudation system?

Denudation rate of 1 to 1.5 m/1000 yr is found in high, rugged mountains. When relief is low (as in E. United States) rate is only about 4 cm/1000 yr.

A MODEL DENUDATION SYSTEM

A MODEL DENUDATION SYSTEM 344

Describe a model denudation system and its progressive changes.

Model denudation system postulates a rapid tectonic uplift of 6 km, taking about 5 million years (m.y.).

What will be the appearance of the region when tectonic uplift has ceased?

Erosion during tectonic uplift has removed much rock and carved the uplifted block into rugged terrain of sharp ridges, deep canyons, and steep slopes.

How is isostatic compensation taken into account in the model denudation system?

Isostatic compensation must be subtracted from gross denudation rate to yield a net value of denudation (assumed to be 0.3 m/1000 yr).

What assumption is made as to the change in denudation rate as time passes?

As denudation continues, rate of denudation diminishes in a constant ratio such that one-half of the available landmass is removed in a fixed time interval, the *half-life*. (Half-life of 15 m.y. is assumed.)

Define *half-life*.

Half-life: time required for an initial quantity at time-zero to be reduced by one-half in an *exponential decay* system.

What is *exponential decay?*

Exponential decay: program of decrease in a variable quantity at a rate such that the quantity is halved in a constant time period, the half-life.

Why does the denudation rate diminish with time?

Denudation diminishes with time as surface elevation is lowered because erosion becomes less and less intense as relief and average slope diminish and stream gradients decline; available potential energy of system is steadily declining.

When is a peneplain produced?

According to terms of model system, peneplain is present when average elevation is 300 m or less, requiring about 60 m.y.

LANDMASS REJUVENATION

LANDMASS REJUVENATION 346

Why does the simple denudation model described above need to be modified?

To approach real geologic conditions, the above denudation model must be modified to include spasmodic uplift, rather than uniform uplift, in isostatic compensation.

What effect has spasmodic uplift on the landscape?

Spasmodic uplift brings sudden large increase in available landmass when average elevation of region is low.

What is *landmass rejuvenation?*

Landmass rejuvenation: episode of rapid fluvial denudation set off by a rapid crustal rise, increasing the available landmass.

Describe the changes that accompany landmass rejuvenation.

Streams undergo rejuvenation first, deeply carving up the uplifted peneplain surface (stage of youth). Steep slopes of stream valleys replace gentle upland slopes of peneplain; region then attains maximum relief and ruggedness (early mature stage). Relief and slopes then gradually decline (late maturity); eventually a new peneplain is formed (old age stage), completing the rejuvenation cycle.

To what climatic regions does the rejuvenation cycle apply?

Rejuvenation cycle applies to moist climate regions in which surplus water flows as stream runoff to reach the sea.

CLIMATE AND LANDSCAPE REGIONS

CLIMATE AND LANDSCAPE REGIONS 347

What is meany by *climogenetic regions?*

Climogenetic regions: concept of a classification of the global land surfaces into several basic landscape types, each having a set of landforms uniquely dictated by climate. (Also called *morphogenetic regions*.)

In what respects is this concept unrealistic?

Concept not fully realistic because geologic controls (rock type, rock structure) strongly influence landforms, in combination with climate.

CLIMATE-PROCESS SYSTEMS

CLIMATE-PROCESS SYSTEMS 347

What is meant by *climate-process systems?*

Climate-process systems: concept of a number of geomorphic systems, each of which contains a unique combination of levels of intensity of several basic denudation processes.

Describe the *periglacial* climate-process system.

The *periglacial* climate-process system, for example, consists dominantly of processes of physical weathering and mass wasting in a cold environment.

What other kinds of climate-process systems may be recognized?

Other possible climate-process systems are those of desert climate, tropical wet-dry climate, steppe climate, and tundra climate.

THE DESERT CLIMATE-PROCESS SYSTEM

THE DESERT CLIMATE-PROCESS SYSTEM 348

Geomorphic Processes of the Desert Environment

Geomorphic Processes of the Desert Environment 348

Describe geomorphic processes in the tropical desert.

Salt-crystal weathering is important, possibly also rock disruption by heating and cooling in diurnal cycle. Hydrolysis and oxidation are active, despite aridity.

What is *desert varnish?*

Desert varnish: dark, iridescent coating found on rock surfaces in a desert climate.

Describe the role of evaporation in desert processes.

Evaporation of soil water produces carbonate deposition in soil as petrocalcic horizon in Aridisols. When exposed at surface, this becomes *calcrete (caliche)* crust.

What is *calcrete?*

Calcrete: rocklike layer rich in calcium carbonate, formed below the soil solum in the C horizon; also called *caliche* in S. W. United States.

Fluvial Processes in an Arid Climate

Fluvial Processes in an Arid Climate 349

What concept relates fluvial process with desert overland and channel flow?

Fluvial processes act effectively in deserts on brief occasions when torrential rains produce overland and channel flow.

How is stream discharge affected by flow in braided channels over alluvial deposits?

Braided streams of arid regions lose discharge by seepage into porous alluvial materials; streams tend to disappear on alluvial deposits or on dry lake floors.

Alluvial Fans

Alluvial Fans 349

What is an *alluvial fan?*

Alluvial fan: a low, gently sloping, conical accumulation of coarse alluvium deposited by an aggrading stream over a piedmont plain below the point of emergence of the channel from a rock-walled canyon. Fans are built as the braided channel shifts radially from fixed apex. Mudflows are commonly interbedded with stream deposits in large desert fans.

How is ground water held beneath an alluvial fan? What role is played by mudflow layers? How can artesian flow occur?

Ground water and alluvial fans: ground water is held in alluvial aquifers capped by impervious mudflow layers, causing artesian flow conditions.

How important are alluvial fans as water reservoirs? What harmful side effects can occur from water overdrafts?

Fans are dominant ground-water reservoirs of southwestern United States. Depletion rapidly occurs by overdrafts fro irrigation. Land surface subsidence is a harmful side effect.

Landscape Evolution in the Basin and Range Province

Landscape Evolution in the Basin and Range Province 350

What U.S. region exemplifies a mountainous desert? What is the origin of the mountain blocks and basins? What landforms are dominant in such a region?

Basin-and-range region of western United States consists of fault block mountains and down-dropped basins. Fans are built into basins, filling them with alluvium. As mountains waste away, pediments develop at the foot of the mountain fronts.

What is a *playa?*

Playa: flat surface underlain by fine sediment or evaporite minerals deposited from shallow lake waters in a dry climate in the floor of a closed depression.

How can alluvial fans create closed basins?

Alluvial fans, built entirely across a desert lowland, can block natural drainage channels and form shallow depressions in which playas form.

What is a *pediment?*

Pediment: a gently sloping rock-floored desert surface lying at the foot of the steep retreating mountain front. Note that a pediment is an erosional landform, whereas an alluvial fan is a depositional landform.

What is a *bajada?*

The Spanish term *bajada* refers to the sloping surface formed of either alluvial fans or pediments; S.W. United States, N. Mexico.

What is an *inselberg?*

Inselberg: a small, islandlike hill or mountain rising sharply above a surrounding pediment or alluvial fan. (German term, means "island mountain.")

Describe the advanced stage of denudation in a mountainous desert. What is a *pediplain?*

In advanced denudation stage of mountainous desert, mountains have been reduced to small remnants or inselbergs, or replaced entirely by pediments. Land surface is largely a composite of fans, pediments, and playas, called a *pediplain.*

Compare a pediplain with a peneplain.

Pediplain is not related to sea level as a base level, as is a peneplain, because pediplain does not have streams leading to the sea.

HILLSLOPE EVOLUTION IN HUMID AND ARID CLIMATES

HILLSLOPE EVOLUTION IN HUMID AND ARID CLIMATES 352

What is a *hillslope?*

Hillslope: a sloping land surface between a drainage divide and a stream channel on a landmass undergoing fluvial denudation.

How is the *hillslope profile* related to the path of overland flow?

Hillslope profile: profile of a hillslope plotted along the path followed by overland flow from a drainage divide to the nearest stream channel.

Why is it difficult to interpret the evolution of hillslopes through time?

Because hillslope changes are often imperceptible over the time period of human observation, the evolutionary patterns of change are difficult to interpret.

What bedrock conditions are assumed in setting up models of hillslope evolution?

Bedrock conditions assumed in hillslope evolution models require that the bedrock be uniform in composition and structure throughout, so that rock structure does not control the profile form.

What simple model can be postulated for hillslope evolution in a humid climate?

In a humid climate, divides will be smoothly rounded because of soil creep and rain-splash erosion; remainder of slope will be straight in profile down to stream channel, where all eroded material is removed as fast as it reaches the slope base.

How will the hillslope profile form change in later stages?

As time passes, hillslope profile declines in steepness; it may develop a concavity at the base, making S-shaped (sigmoid) profile form.

Describe hillslope evolution in an arid climate.

In arid climate, hillslope is postulated to retain constant angle, to retreat at base as pediment develops and widens. Ultimate form is pediment surface.

What is *reclining retreat?*

Reclining retreat: program of hillslope evolution in which the angle of slope decreases with time; as postulated for a humid climate.

What is *parallel retreat?*

Parallel retreat: program of hillslope evolution in which the angle of slope remains essentially constant as the slope retreats.

THE CLIMATE-PROCESS SYSTEM OF THE SAVANNA ENVIRONMENT

THE CLIMATE-PROCESS SYSTEM OF THE SAVANNA ENVIRONMENT 353

Describe the geomorphic processes of the savanna environment.

Geomorphic processes: chemical weathering may be important in wet season; deep regolith may be present. Some areas show *bornhardts* and *tors*, with laterite crusts surrounding them.

What is a *bornhardt?*

Bornhardt: prominent rock knob of massive granite or similar plutonic rock with rounded summit and often showing exfoliation shells.

What is a *tor?*

Tor: group of boulders or joint blocks forming a small, but conspicuous hill.

Describe *laterite (ferricrete)* crusts.

Land surfaces may show capping of *laterite (ferricrete)* forming hard crust over regolith.

What is *laterite?*

Laterite: rocklike layer rich in sesquioxides of aluminum and iron, including the minerals bauxite and limonite, found in low latitudes in association with Ultisols and Oxisols. May represent plinthite layer that has become hardened upon exposure to surface.

What is *ferricrete?*

Ferricrete: alternate term for laterite; used in context of various kinds of hard surface crusts, such as silcrete and calcrete.

What is the significance of crusts of ferricrete and silcrete in terms of past climate?

Laterite (ferricrete) and silcrete may have originated as soil horizons in earlier, moister climate; have since been uncovered by erosion.

What is *silcrete?*

Silcrete: crust cemented largely by silica.

CHAPTER 18—SAMPLE OBJECTIVE TEST QUESTIONS

A. MATCHING

1.	tectonic	a. _______	uplift
2.	half-life	b. _______	evaporites
3.	rejuvenation	c. _______	islandlike
4.	playa	d. _______	crustal breakage
5.	alluvial fan	e. _______	parallel retreat
6.	inselberg	f. _______	braided stream
7.	hillslope	g. _______	exponential decay

B. MULTIPLE CHOICE

1. Because of isostatic compensation, the available landmass is

   ________a. decreased.

   ________b. increased.

   ________c. alternately raised and lowered.

   ________d. lowered beneath the sea.

2. Compared with the rate of landmass denudation by fluvial processes, the rate of tectonic uplift is

   ________a. about the same.

   ________b. slightly slower.

   ________c. much more rapid.

   ________d. much slower.

3. A pediment is formed

   ________a. in an arid climate.

   ________b. in a humid climate.

   ________c. by erosion processes.

   ________d. (both a. and c., above.)

C. COMPLETION

1. The gently undulating land surface of low elevation, formed at the close of the fluvial denudation cycle in a humid climate is a ______________________________.

2. The principle of flotation of the earth's crust in equilibrium is called ______________________________.

3. Uplift of a landmass after the denudation cycle is largely completed leads to ______________ ________________________ of the landmass.

4. A flat desert surface produced by sediment deposition in a shallow lake is a ______________ ________________________.

5. A land surface produced at the close of a denudation cycle in a mountainous desert is a _____ ________________________.

CHAPTER 19

LANDFORMS AND ROCK STRUCTURE

PAGE

What concept relates crustal uplift to denudation?

Concept: crustal uplift increases the thickness of landmass available to denudation, and in this way deep-seated rocks and structures appear at the surface.

State the concept of control of landforms by rock structure. How is rate of denudation controlled by rock type?

Concept: sequential landforms of the erosional class are controlled in shape, size, and arrangement by the underlying rock structure. Denudation acts more rapidly upon weaker rock types, lowering them to produce valleys; stronger rocks stand out as ridges and uplands.

EROSIONAL LANDFORMS AND CONTINENTAL HISTORY

EROSIONAL LANDFORMS AND CONTINENTAL HISTORY 355

Which erosional landforms are described in this chapter?

Described in this chapter are erosional landforms developed on the stable crust of the continental lithosphere.

What is the general history of stable continental shields since the close of the Cretaceous Period?

Since the close of the Cretaceous Period, continental shields have experienced fluvial denudation, with some up-and-down-warping. Sporadic volcanic activity has occurred as shields have moved over mantle plumes.

ROCK STRUCTURE AS A LANDFORM CONTROL

ROCK STRUCTURE AS A LANDFORM CONTROL 355

In what way do sedimentary strata differ in resistance to erosion?

Sedimentary strata differ in resistance to erosion, giving rise to a variety of landforms.

What landforms are associated with shale? Why?

Shale is weak rock, easily eroded to form valleys.

What landforms are associated with limestone? In a humid climate? In an arid climate?

Limestone, susceptible to carbonic acid action, forms valleys in humid climate. Stands high in arid climate, forming summits or cliffs.

What landforms are made by sandstone and conglomerate?

Sandstone and conglomerate are resistant, form ridges.

What features are associated with igneous rocks? With metamorphic rocks?

Igneous rock usually forms uplands with respect to sedimentary strata. Metamorphic rocks are more resistant than sedimentary parent types, but show differences among types.

DIP AND STRIKE

DIP AND STRIKE 356

How is the attitude of a natural rock plane described?

Geometrical description of attitude of natural rock planes used dip and strike.

Define *dip*. What units of measure are used?

Dip: acute angle formed between rock plane and horizontal plane. Measured in degrees: horizontal is 0°; vertical is 90°.

Define *strike*.

Strike: compass direction of line of intersection between inclined rock plane and horizontal plane.

STRUCTURAL GROUPS OF LANDMASSES

STRUCTURAL GROUPS OF LANDMASSES 356

Name three major groups of landmasses.

Landmasses fall into three major groups:

A. Undisturbed sedimentary strata.

B. Disturbed structures of tectonic activity.

C. Eroded igneous masses.

What two types of landmasses are included in Group A, the undisturbed sedimentary strata? Where are thick sedimentary strata found?

Group A. Undisturbed sedimentary strata. Thick covers of sedimentary rocks overlying ancient shield rocks. Strata lie nearly horizontal. Two types are (a) horizontal strata, (b) coastal plains.

What are the four types of disturbed structures of tectonic activity?

Group B. Disturbed structures of tectonic activity show bending and breaking of crust. Four types are (a) domes, (b) folds, (c) fault structures, (d) metamorphic belts.

What are the two basic types of eroded igneous masses?

Group C. Eroded igneous masses. Two basic types: (a) exposed batholiths; (b) extinct volcanoes.

UNDISTURBED SEDIMENTARY STRATA

UNDISTURBED SEDIMENTARY STRATA 357

Where are undisturbed sedimentary strata found? What is the age of these strata?

Undisturbed sedimentary strata, more-or-less horizontal, cover large parts of the continental shields. Strata range in age from Paleozoic through Cenozoic.

What are the two major classes of undisturbed strata?

Two major classes of undisturbed strata: interior shield covers and coastal plains.

INTERIOR SHIELD COVERS

INTERIOR SHIELD COVERS 357

Where do horizontal strata occur? Of what geologic age are the strata? How did these strata come to be elevated?

Covers on continental shields consist of horizontal strata, younger than Precambrian age. Most are marine strata, deposited in shallow seas, then uplifted.

What are the landforms of distinctive shape developed on horizontal strata? Under what climate are they best shown? Give an example of a region of horizontal strata.

Landforms of horizontal strata include cliffs, plateaus, canyons, mesas, and buttes. Arid climate allows sharp definition of these features. Example: Colorado Plateau region.

What is a *cliff*? What kinds of rock form cliffs? What process maintains the steepness of a cliff?

Cliff: sheer, nearly vertical rock wall formed from flat-lying hard rock formation, usually sandstone or limestone. Continual undermining of cliff is accompanied by breakoff of rock masses along joints.

What is a *plateau* of horizontal strata?

Plateau: upland surface, more or less flat and horizontal, upheld by a resistant bed or formation of sedimentary strata or lava flows and bounded by a steep cliff.

What is a *mesa?* What form has a mesa? How does it change with continued erosion?

Mesa: table-topped plateau remnant bordered on all sides by cliffs. Mesa is remnant of a resistant rock layer; shrinks in area as cliffs retreat.

What is a *butte?* What stage in erosion of horizontal strata does a butte represent.

Butte: prominent, steep-sided hill or peak, often representing the final remnant of resistant rock layer as mesa is consumed.

What kind of drainage pattern develops on horizontal strata?

Stream network on horizontal strata takes the dendritic pattern.

Describe a *dendritic drainage pattern.*

Dendritic drainage pattern: treelike branched stream pattern in which smaller streams take a wide variety of directions. Parallelism of streams is lacking. (Also occurs in areas of plutonic rock.)

Can flood basalts produce landforms similar to sedimentary strata? Give an example.

Flood basalts produce landforms resembling those of horizontal strata. Examples: Columbia Plateau, Deccan Plateau.

LIMESTONE CAVERNS

LIMESTONE CAVERNS 359

What are *caverns?* How are caverns formed? In what kind of rock? How is cavern formation related to the ground water table? How are cave deposits formed?

Caverns: interconnected subterranean cavities in limestone, formed by dissolving action of circulating ground water. Removal is concentrated in upper part of ground water zone.

What is a *sinkhole?* With what systems are sinkholes connected?

Sinkhole: a surface depression in limestone, connected to caverns beneath.

What is *dripstone?*

Dripstone: encrustations forming where water drips from cave ceilings.

What is *flowstone?*

Flowstone: encrustations made in moving water pools and streams on cave floors.

What are *stalactites?*

Stalactites: scenic feature of caverns where slow drip of water from the ceiling creates spikelike forms built downward.

What are *stalagmites?*

Stalagmites: scenic feature of cavern floor upon which steady drip of water forms postlike columns built upward.

Describe some environmental aspects of caverns. How have humans used caverns? What useful commodity comes from certain caverns?

Environmental aspects of caverns: important as habitations for early humans; used today for storage, living space, industry. Bats living in caverns accumulate guano, a deposit rich in nitrates, used as fertilizer.

KARST LANDSCAPES 361

What is *karst?* What surface features are abundant in a karst region? Are surface streams present? Name two karst regions.

Karst: a landscape or topography dominated by surface features of limestone solution and underlain by a cavern system. Sinkholes are abundant; surface streams nonexistent. Examples: Mammoth Cave region of Kentucky, Yucatan peninsula.

Define *cockpit karst.*

Cockpit karst: a landscape consisting of numerous closely-set hills with rounded summits; between them are closed depressions with sinkholes (found in Puerto Rico).

Define *tower karst.*

Tower karst: a landscape of steep sided conical, limestone hills 100–500 m (300–1500 ft) high (found in southern China and W. Malaysia).

What are "mogotes?"

"Mogotes": are isolated, limestone tower karst knobs found in Puerto Rico.

ENVIRONMENTAL AND RESOURCE ASPECTS OF SHIELD COVERS 362

What rocks of horizontal strata have economic value? How are they used? Do sedimentary rocks contain metallic ores?

Rocks of economic value include building stone, limestone for portland cement and iron smelting; some strata contain metallic ores such as lead, zinc, uranium.

Describe fossil fuel resources of sedimentary strata. How will occurrence of coal seams in sedimentary strata invite strip mining environmental impact?

Coal and petroleum are major fossil fuel resources of sedimentary strata. Large reserves of coal lie close to surface in horizontal strata of western U.S., inviting strip mining. Scarification and pollution anticipated as environmental impact of strip mining.

Why has the role of fossil fuels come under question recently?

Coal, lignite, petroleum and gas all cause an increase in atmospheric carbon dioxide during combustion which is a source of concern in current problem of global warming.

Describe the ways in which deep coal seams are mined. What environmental damage results?

Deep coal seams are reached by vertical shafts and horizontal tunnels (*drifts*). Abandoned workings collapse, causing surface subsidence of land.

What method is used to extract coal close to the surface?

Coal seams close to surface are worked by *strip mining* in which overburden is removed and coal extracted by power machines. Two kinds of strip mining are *area strip mining* and *contour strip mining.*

Define *area strip mining.*

Area strip mining: form of strip mining practiced in regions where a horizontal coal seam lies beneath a land surface that is approximately horizontal. Overburden is heaped into parallel spoil ridges.

Define *contour strip mining.*

Contour strip mining: form of strip mining practiced in hilly or mountainous regions where coal seams form natural outcrops along the contour of the hillslopes. Overburden is dumped in terrace on downhill side; a clifflike high wall remains on uphill side.

COASTAL PLAINS

COASTAL PLAINS 364

What is a *coastal plain?* Of what kind of rock is a coastal plain composed? What is the age of most coastal plain strata?

Coastal plain: a coastal zone, emerged from beneath sea as a former continental shelf, underlain by strata dipping gently seaward. Most coastal plain strata are geologically young, of Cenozoic age.

Interpret coastal plains in the light of plate tectonics.

Coastal plain sediment wedges form on subsiding margin of continental lithosphere, where bordered by oceanic lithosphere and not on an active plate boundary.

What causes subsidence of the continental shelf?

Subsidence is caused by loading of crust with sediment, under principle of isostasy.

What kind of crustal movements have affected the coastal plain during its development?

Epeirogenic movements, both uplift and down-sinking, have affected the coastal plain belt during its history.

Describe the drainage that first appears on a coastal plain. What class of streams forms first? What class of streams is formed later? What drainage pattern develops?

Drainage on coastal plain first consists of seaward-flowing *consequent stream;* later, *subsequent streams* develop on *lowlands* between *cuestas* to form a *trellis drainage pattern.*

What is a *consequent stream?* In what direction do consequent streams flow on a coastal plain?

Consequent stream: a stream that takes its course down the slope of an initial land surface. On coastal plain, consequent streams flow across belt of newly exposed strata to reach sea.

What is a *lowland?* What kind of rock underlies a lowland? What kind of streams occupy lowlands?

Lowland: a broad valley eroded upon weak strata, usually clays, and trending parallel with the coastline. Lowlands are occupied by subsequent streams.

What is a *cuesta?*

Cuesta: erosional landform developed on resistant strata of low to moderate dip and taking the form of an asymmetrical low ridge or hill belt with one side a steep scarp or descent and the other a gentle slope.

What is the *inner lowland?*

Inner lowland: the lowland lying between an area of older rock (*oldland*) and the first cuesta of a coastal plain.

Define *subsequent stream.* In what kinds of rocks and on what structures can subsequent streams be found?

Subsequent stream: a stream that develops its course by erosion of a band or belt of weaker rock. Subsequent streams form on many kinds of structures of dipping strata, and along faults.

Describe a *trellis drainage pattern.* How are the principal streams arranged? Which streams are consequent? Which are subsequent?

Trellis drainage pattern: a stream network in which the principal streams form two sets oriented at about right angles to one another. In the case of the coastal plain, one set of streams is consequent in origin, the other set is subsequent. Other types of trellis patterns exist (see folds).

COASTAL PLAIN OF NORTH AMERICA

COASTAL PLAIN OF NORTH AMERICA 365

Describe the coastal plain of the United States. Give width and length. Where does it start at the north end? How far along the coast does it extend?

Coastal plain of U.S. is a major geographical region; a belt 160–500 km (100–300 mi) wide and 3000 km (2000 mi) long. Starts at Long Island, N.Y., continues south and west along Gulf Coast to Mexico.

ENVIRONMENTAL AND RESOURCE ASPECTS OF COASTAL PLAINS

ENVIRONMENTAL AND RESOURCE ASPECTS OF COASTAL PLAINS 365

Where is agriculture important on a coastal plain? What economic value have cuestas?

Agricultural value often high on broad fertile lowlands. Cuestas bear forests (pine forests in U.S.).

In what way are transportation routes influenced by a coastal plain? Give an example.

Transportation lines run parallel with coastline, following lowlands. Example: New York City to Richmond is chain of cities connected by major rail and highway lines.

What is Fall Line? What is its significance: How has it influenced urbanization?

Fall Line: inner limit of coastal plain strata of Atlantic seaboard. Tidal streams of coastal plain navigable to this point, where falls and rapids begin. Site of development of major U.S. cities.

Explain how artesian water supplies can occur beneath coastal plains. What material forms the aquifer? the aquiclude?

Artesian water supplies important in coastal plain. Sandy cuestas absorb precipitation. Clay aquicludes confine water in aquifers.

What fossil fuel resources are important beneath coastal plains?

Petroleum and natural gas resources are important in broad, coastal plains of thick strata, also offshore beneath shelf.

Name several mineral resources of coastal plains.

Other mineral resources: lignite, sulfur, phosphate beds, ceramic clays.

SEDIMENTARY DOMES

SEDIMENTARY DOMES 366

What is a *sedimentary dome?* In what parts of the continents are such domes found?

Sedimentary dome: uparched strata forming a circular structure with domed summit. Occurs in sedimentary strata of covered shields.

Describe the features of erosion of a sedimentary dome. What kinds of valleys are developed? What lies in the dome center?

Erosion features of domes include hogbacks, annular (ringlike) subsequent valleys, core of ancient shield rocks.

Describe a *hogback*. What kinds of rocks form hogbacks? What pattern do hogbacks make within a dome?

Hogback: a sharp-crested, often sawtooth ridge formed of the upturned edge of resistant rock layer. Hogbacks curve in circular pattern around flanks of eroded dome.

Describe an *annular drainage pattern*. What class of streams makes up the pattern?

Annular drainage pattern: stream network dominated by concentric (ringlike) major subsequent streams. Shorter tributaries of radial pattern complete the network.

THE BLACK HILLS DOME

THE BLACK HILLS DOME 367

Use the Black Hills dome as example of a large, partly eroded dome. Describe the landforms in detail. How have these landforms influenced transportation and placement of cities?

Black Hills dome of western South Dakota and eastern Wyoming is prime example of large, flat-topped dome structure, partly eroded. Red Valley, called Race Track, is subsequent valley circling entire dome. Highways and railroads connect towns located in Red Valley.

What rock type comprises the exposed core of the Black Hills dome? What value has this central area? What mineral deposits does it contain?

Core of Black Hills dome has exposed intrusive and metamorphic rocks. Mountainous terrain with forests and parks. Valuable metallic ore deposits, including gold of Homestake Mine.

LANDFORMS OF ERODED FOLD BELTS

LANDFORMS OF ERODED FOLD BELTS 367

Explain how folded strata give rise to landforms reflecting the structure.

Alternating resistant and weak strata in fold sequence of anticlines and synclines give rise to parallel ridges and valleys during erosion of fold belt.

What is a *syncline?*

Syncline: a troughlike downfold of strata.

What is an *anticline?* What is the relationship of an anticline to a syncline?

Anticline: an archlike upbend of strata. Anticlines alternate with synclines in fold belts.

What varieties of valleys and mountains are developed by erosion of open folds?

Erosion of open folds produces *synclinal valleys* and *anticlinal mountains,* followed by reversal of topography to form *anticlinal valleys* and *synclinal mountains.*

What is a *watergap?*

Watergap: narrow transverse gorge cut across a narrow ridge by a stream, usually in a region of folds.

What is an *antecedent stream?*

Antecedent stream: stream that has maintained its course across a rising rock barrier, such as an anticlinal fold or a fault block.

What drainage pattern forms on eroded folded strata? How does this pattern differ from the trellis pattern of a coastal plain?

Trellis drainage pattern on folds consists of long parallel subsequent streams with many short tributaries at right angles.

ZIGZAG RIDGES AND PLUNGING FOLDS

ZIGZAG RIDGES AND PLUNGING FOLDS 369

What are *plunging folds?*

Plunging folds: folds of strata with descending or rising crests (fold axes). When eroded, they give rise to zigzag lines of ridges, formed of alternate plunging synclines and anticlines.

ENVIRONMENTAL AND RESOURCE ASPECTS OF FOLD REGIONS

ENVIRONMENTAL AND RESOURCE ASPECTS OF FOLD REGIONS 371

Describe the Appalachian Mountains of the eastern United States as an example of eroded folds in a region of humid climate. Describe the landscape. How high are the ridges? How are transportation lines influenced?

Appalachian Mountains of eastern U.S. are good example of fold belt in humid climate. Illustrations drawn from eastern Pennsylvania, where conglomerate ridges rise 600 m (2000 ft) above valley floors. Highways follow valleys, cross from one valley to next through watergaps. Ridges are barriers to communication.

What form of coal occurs in folded strata? In what structures is it preserved?

Anthracite coal is often found in folded strata; seams lie deep down in synclines not yet eroded away.

How does petroleum occur in folds? In what structure does it occur?

Petroleum occurs in fold traps. Oil migrates to crest of anticline, held in sandstone. Oil also accumulates in similar manner in sedimentary domes.

EROSIONAL DEVELOPMENT OF FAULT SCARPS AND BLOCKS

EROSIONAL DEVELOPMENT OF FAULT SCARPS AND BLOCKS 372

How does long-continued erosion sustain distinctive landforms along a fault? Why does evidence of the fault persist?

Fault scarp is eroded away, but fault-controlled landform persists because of weakness of crushed rock on fault plane, extending deep down into crust.

What is a *fault-line scarp?* Why does the scarp persist? What type of stream is often located on the fault line? Explain.

Fault-line scarp: a long, nearly straight scarp developed by long-continued erosion along a fault line. A stream may run parallel at base of scarp, eroding crushed zone of fault.

Describe an eroded tilted fault block. What vestiges of the fault plane persist? What is the name for these features?

Erosion of tilted fault block produces rugged mountain mass. Fault plane may continue to appear as line of *triangular facets.*

What are *triangular facets?*

Triangular facets: steeply inclined bedrock surfaces of triangular outline occurring between canyons carved into the fault scarp near the base of a fault block mountain.

ENVIRONMENTAL RESOURCE ASPECTS OF FAULTS AND BLOCK MOUNTAINS

ENVIRONMENTAL AND RESOURCE ASPECTS OF FAULTS AND BLOCK MOUNTAINS 373

How are ore deposits related to fault planes? Explain.

Ore deposits may occur along fault planes, where ore-forming solutions rise and deposit minerals.

How are springs related to fault planes? Explain. Give one example.

Springs, both hot and cold, often issue from fault lines. Examples: Arrowhead Springs, Palm Springs in California.

How is petroleum trapped by fault structures?

Petroleum traps formed by faults cutting across strata.

In what way do fault scarps and fault-line scarps form barriers? Give one example.

Fault scarps and fault-line scarps can form imposing topographic barriers. Example: Hurricane Ledge, Utah.

What major landforms are associated with grabens? Give one example.

Grabens may form broad lowlands. Example: Rhine Graben, Germany.

METAMORPHIC BELTS

METAMORPHIC BELTS 373

What landforms develop by erosion of metamorphic rocks? Which rock types form ridges? Which form valleys? Give an example.

Erosion of metamorphic rocks usually gives long ridges and valleys in parallel pattern, but not as sharply defined as folded strata. Ridges develop on quartzite, hill belts on slate and schist, lowlands on marble. Example: Taconic and Green mountains of New England.

EXPOSED BATHOLITHS

EXPOSED BATHOLITHS 374

Where are exposed batholiths found? Give one example.

Batholiths are uncovered by erosion; make up large parts of continental shields. Example: Idaho batholith, Sierra batholith.

What drainage pattern develops on plutons? How does this pattern compare with that developed on eroded horizontal strata?

Dendritic drainage pattern develops on plutons. Not easily distinguished from pattern on horizontal strata.

What is a *monadnock?*

Monadnock: prominent isolated mountain or large hill rising conspicuously above a surrounding peneplain and composed of a rock more resistant than that which underlies the peneplain.

EROSIONAL LANDFORMS OF VOLCANOES AND LAVA FLOWS

EROSIONAL LANDFORMS OF VOLCANOES AND LAVA FLOWS 376

Describe the stages in erosion of composite volcanoes. How does the external form change? What drainage pattern results?

Extinct composite volcanoes are dissected by stream erosion, obliterating the crater or caldera and destroying the smooth conical form. Radial drainage pattern results.

What is a *radial drainage pattern?*

Radial drainage pattern: pattern of streams radiating outward from central point.

What landforms are developed on eroded lava flows?

Lava flows may form high mesas (flat-topped hills) as weaker rock is eroded away from adjoining areas.

What landforms emerge in the final stages of volcano erosion?

In advanced stage solidified lava in feeder pipe of volcano forms narrow, sharp peak, a volcanic neck, while radial dikes stand in relief as walls.

What is a *volcanic neck?* What form has a volcanic neck? Give an example.

Volcanic neck: isolated, narrow, steep-sided peak formed of igneous rock previously solidified in feeder pipe of volcano. Example: Ship Rock, New Mexico.

Describe the landforms produced by erosion of Hawaiian shield volcanoes.

Shield volcanoes of Hawaii are carved by deep, steep-walled canyons; these enlarge and multiply until rugged mountain mass is formed.

CHAPTER 19—SAMPLE OBJECTIVE TEST QUESTIONS

A. MATCHING

1.	shale	a. _______	downfold
2.	strike	b. _______	ringlike
3.	Fall Line	c. _______	batholith
4.	mesa	d. _______	flat-top rock
5.	anticline	e. _______	steep dip
6.	syncline	f. _______	coastal plain
7.	hogback	g. _______	compass direction
8.	annular	h. _______	upfold
9.	pluton	i. _______	weak rock

B. MULTIPLE CHOICE

1. A consequent stream is one that

________a. occupies a zone of weak rock.

________b. runs along the lowland of a coastal plain.

________c. always has a steep gradient.

________d. flows down the slope of a new land surface.

2. A trellis drainage pattern

________a. is typical of a belt of eroded folds.

________b. is formed during erosion of a volcano.

________c. has a treelike branching pattern.

________d. consists of consequent streams.

3. On a map, an eroded sedimentary dome can be distinguished from an eroded composite volcano by the presence of

________a. a caldera.

________b. zigzag folds.

________c. annular drainage pattern.

________d. dendritic drainage pattern.

C. COMPLETION

1. The acute angle formed between a natural rock plane and the horizontal plane is called the __ ________________________.

2. Eroded edges of steeply dipping strata form sharp-crested saw-tooth ridges called __________ ________________________.

3. A treeline branching stream system makes up the ______________________________ drainage pattern.
4. A type of stream that maintains its course across a rising barrier, such as a fold or fault block, is an ______________________________ stream.

CHAPTER 20

LANDFORMS MADE BY WAVES AND CURRENTS

PAGE

WAVES IN DEEP WATER

WAVES IN DEEP WATER 378

What is the energy source of waves of oceans and lakes?

Waves of oceans and lakes are generated almost entirely by winds, which transfer energy from atmosphere to ocean surface.

What are *oscillatory waves?*

Oscillatory waves: water waves in which particles move in vertical orbits, completing one orbital circle with the passage of one wave.

What terms are used to describe water waves?

Water waves are described in terms of wave height and wave length.

SHOALING WAVES AND BREAKERS

SHOALING WAVES AND BREAKERS 379

At what depth do shoaling waves encounter interference with the bottom?

Shoaling waves encounter interference with the bottom at depth equal to about one-half the wave length. Friction develops with bottom.

In what way is the wave form changed?

Shoaling wave slows in speed, increases in height; crest steepens and collapses as a *breaker.*

Define *breaker.* How is the wave motion changed when breaking occurs?

Breaker: wave that has steepened and collapsed upon reaching shallow water. Wave motion of deep water is transformed into strong forward surge.

Describe the *swash (uprush).* What effect has swash on beach sediment?

Swash (uprush): surge of water up the beach slope (landward) following collapse of a breaker. Swash carries sand up beach slope.

Describe the *backwash.* What function does backwash perform?

Backwash: return flow of swash water under influence of gravity. Backwash carries sand seaward down beach slope.

MARINE EROSION OF COASTS

MARINE EROSION OF COASTS

Define *shoreline.*

Shoreline: the shifting line of contact between water and land.

Define *coastline,* or, *coast.* What processes act along coastlines? What zones of activity and what features are included in the coastline?

Coastline, or *coast:* a zone in which coastal processes operate or have a strong influence. Processes include waves, currents, tides, and winds. Coastline includes shallow water zone in which waves act; also beaches, coastal dune belt, and sea cliffs.

What is a *marine cliff?*

Marine cliff: a cliff of rock shaped and maintained by the undermining action of breaking waves. Changes in cliff are extremely slow in hard rock.

Under what conditions is marine erosion rapid?

Marine erosion is rapid in weak sedimentary strata, residual regolith, alluvium, glacial deposits, and sand dunes.

What coastal landform develops where marine erosion is rapid?

In weak materials, a *marine scarp* is formed, with recession rates on the order of 1 m/yr. Example: outer shoreline of Cape Cod.

Describe a *marine scarp.*

Marine scarp: a high coastal bank of poorly consolidated alluvium or other forms of regolith, standing at a slope angle of 30°–35° in the case of sand.

What is *retrogradation?*

Retrogradation: the cutting back (retreat) of a shoreline, beach, marine cliff, or marine scarp by wave action.

What is a *sea cliff?*

Sea cliff: synonym for marine cliff.

What are the common landforms associated with a sea cliff?

Landforms of a sea cliff include a *wave-cut notch, crevices, sea caves, arches, stacks, an abrasion platform, and hanging valleys.*

What is a *wave-cut notch?*

Wave-cut notch: recess at the base of a marine cliff (sea cliff) where wave impact is concentrated.

What are *crevices?*

Crevices: narrow, cleftlife cavities eroded into the face of a sea cliff by breaking storm waves.

What are *sea caves?*

Sea cave: cave near the base of a sea cliff, eroded by breaking waves.

What are *arches?*

Arch: arch of bedrock remaining when a rock promontory of a sea cliff has been cut through on two sides by wave action.

What is a *stack?*

Stack: isolated, columnar mass of bedrock left standing during retrogradation of a sea cliff.

What is an *abrasion platform?*

Abrasion platform: sloping, nearly flat bedrock surface extending seaward from the foot of a marine cliff under shallow water of the breaker zone; it is formed by abrasion of gravel and cobblestones moved by the swash and backwash.

What is a *hanging valley?*

Hanging valley: stream valley that has been truncated by marine erosion so as to appear in cross section in the upper part of a marine cliff. (Also produced by glacial erosion in upper wall of a glacial trough.)

What value have sea cliffs?

Sea cliffs have value as scenic features for recreational use, and as habitats for specialized ecosystems.

BEACHES

BEACHES 381

What is a *beach?* Of what materials is a beach formed?

Beach: a thick, wedge-shaped accumulation of sand, gravel, or cobbles in the zone of breaking waves.

What is the concept of beach function in terms of wave energy? When are beaches built outward?

Beach absorbs energy of storm waves, which cut away the beach. Beach is gradually rebuilt during periods of low waves.

Describe the sorting of sediment in beaches. How is slope angle of beach related to particle size?

Beach sediment, ranging in size from fine sand to coarse gravel and cobblestones, is well sorted by wave action. Beaches of fine sand have a gentle slope, those of cobbles have a steep slope with a high crest or ridge.

What happens to silt and clay particles eroded by waves?

Silt and clay particles eroded by waves are carried seaward in suspension, to be deposited in deeper water.

THE BEACH PROFILE

THE BEACH PROFILE 382

What features will be found in the profile of a typical beach in summer in midlatitudes?

Typical features of a beach profile in summer in midlatitudes include the following: *summer berm, winter berm, foreshore, offshore bar, offshore zone.*

What is the *summer berm?*

Summer berm: benchlike sand structure built by swash during the summer season.

What is the *winter berm?*

Winter berm: high berm on landward side of summer berm, built by high waves of winter season.

What is the *foreshore?*

Foreshore: sloping face of a beach that is within the zone of swash and backwash, i.e., alternately covered with water and exposed.

What is an *offshore bar?*

Offshore bar: a low bar of sand and gravel in the offshore zone of a beach.

What is the *offshore?*

Offshore: that part of a beach lying in the zone of shoaling waves and below the level of low tide.

What is *progradation?* What causes it to occur?

Progradation: shoreward building of a beach, bar, or sandspit by addition of coarse sediment carried by littoral drift or brought from deeper water offshore.

What are *beach ridges?*

Beach ridges: low coastal sand or pebble ridges representing berms produced in succession during progradation of a beach.

What can cause the retrogradation of a beach?

Beach retrogradation can occur when sediment is carried out into deeper water, or moved along shore to another area.

LITTORAL DRIFT

LITTORAL DRIFT 382

Describe the oblique approach of waves to a shoreline.

Most of the time, waves approach a shoreline obliquely. As the waves travel through shallow water the wave front is bent by the process of *wave refraction.*

Define *wave refraction.*

Wave refraction: bending of a wave front as it travels obliquely across a rising bottom slope in shallow water. Bending turns the wave front to closer parallelism with shoreline.

What is the effect of oblique wave approach on swash and backwash?

Oblique wave approach causes swash to ride obliquely up slope of beach; backwash returns in direct downhill path by gravity. Rock particles are carried up the beach obliquely, causing mass transport along the shoreline by *beach drift.*

Define *beach drift.* What is the importance of beach drifting?

Beach drift: transport of beach sand parallel with shoreline at times when swash rides obliquely up the beach.

What is *longshore drift?* What causes longshore drifting?

Longshore drift: movement of sediment on sea floor beneath breakers in a direction parallel with the shore. Caused by a *longshore current.*

What is a *longshore current?*

Longshore current: a current in the breaker zone running parallel with the shoreline, set in motion by oblique approach of waves.

What is *littoral drift?* What actions are combined in littoral drift? What activity is performed by littoral drift? What landforms result from littoral drift?

Littoral drift: transport of sediment parallel with the shoreline by the combined action of beach drifting and longshore drifting. Littoral drift carries sand into open water of bays, resulting in building of *sandspits.*

What is a *sandspit?* How does sandspit growth alter the form of a bay?

Sandspit (spit): narrow, fingerlike embankment of sand constructed by littoral drift into open water of a bay.

What is a *bar?*

Bar: a long, low, narrow sand deposit built by wave action; usually attached to higher coastal land at both ends.

What is a *baymouth bar?*

Baymouth bar: sand bar built above water level and extending across the entire mouth of a bay.

What is a *tombolo?*

Tombolo: sand bar connecting an island with the mainland.

Describe a *cuspate bar.*

Cuspate bar: low coastal beach ridge projecting seaward in a toothlike form; results from littoral drift converging from opposite directions along a shoreline.

What is a *cuspate foreland?*

Cuspate foreland: an accumulation of beach ridges and *swales* projecting seaward in a cuspate or arcuate form, developed by long-continued progradation. Example: Cape Canaveral.

What are *swales?*

Swales: long, narrow depressions lying between beach ridges (or between bars on the inside of a meander bend on a floodplain).

WAVE REFRACTION ON AN EMBAYED COAST

WAVE REFRACTION ON AN EMBAYED COAST 384

Describe wave refraction and energy distribution on an embayed coast.

On embayed coast (bays and headlands) refraction bends waves to converge on headlands; here energy is concentrated and erosion is intense, producing sea cliffs. Waves entering bays are weakened. Oblique approach of waves to sides of bays carries sediment from headland to bay head, producing a *pocket beach.*

What is a pocket beach?

Pocket beach: beach of crescentic outline located at bay head.

LITTORAL DRIFT AND SHORE PROTECTION

LITTORAL DRIFT AND SHORE PROTECTION 385

What two forms of change can a beach experience?

Beaches can experience two forms of changes: progradation and retrogradation.

What are the undesirable effects or retrogradation?

Retrogradation leads to property destruction, including undermining of sea walls and shore properties.

What is the concept of inducing progradation to protect property? Explain fully. What kinds of structures are used to induce progradation?

Induced progradation will result in a wide buffer beach to absorb wave energy in times of storm, thus protecting property. Progradation can be induced by construction of *groins*, trapping littoral drift.

What is a *groin?* Of what materials is a groin constructed? What is the effect of groin? For what reason are multiple groins used?

Groin: a wall or embankment built out into the water at right angles to the shoreline. May be constructed of rock masses, concrete, or wooden pilings. Sand accumulates as beach on up-drift side of groin. Beach is depleted on down-drift side. Multiple groins usually necessary to cause progradation of long stretch of beach.

What effect has the construction of a river dam upon the coastline near the river mouth? Give an example.

Effect of upstream dams is to hold back sediment reaching shoreline, causing retrogradation of beaches. Example: Nile delta.

TIDAL CURRENTS

TIDAL CURRENTS 386

What is the concept of tidal changes of ocean level as a cause of strong currents?

Tidal rise-and-fall of ocean level sets in motion strong currents capable of modifying the coastline.

What is the *ocean tide?* What causes the tide?

Ocean tide: rhythmic or cyclic rise and fall of ocean level under changing gravitational attractive forces of moon and sun.

Describe a *tide curve?* What information does it convey?

Tide curve: a graphic representation of rise and fall of ocean level because of tides. Height scaled on vertical axis; time on horizontal axis.

Define *high water.*

High water: point in time at which water level reaches its highest point during a tidal cycle.

Define *low water.*

Low water: low point of water level in tidal cycle.

What is the *range of tide?*

Range of tide: difference in heights of successive high and low waters.

What are *tidal currents?* How does direction very in tidal currents? Where are tidal currents founds?

Tidal currents: alternate landward and seaward flows of water in bays, inlets, and estuaries in response to rise and fall of tide level in open ocean.

Define *ebb current.* When does the ebb current set in?

Ebb current: seaward flow of tidal currents, setting in when tide begins to fall.

Define *flood current.* When does it set in?

Flood current: landward flow of tidal current, setting in when tide begins to rise.

TIDAL POWER

TIDAL POWER 387

How is *tidal power* harnessed?

Tidal power is harnessed by placing dams with turbines at mouths of bays with large tidal range.

Describe tidal power development by Canada at the Bay of Fundy.

A pilot plant generating tidal power was put into operation in 1984, at Annapolis Royal, Nova Scotia. It generates about 20 megawatts of electricity by using the ebb flow through a dam placed at the entrance of a long, narrow bay. Other sites are being studied in the Bay of Fundy.

What studies have been made in Great Britain about utilizing tidal power?

In Great Britain, studies have been made to use large, shallow bays for tidal power, Long, low dams would be built offshore with many sluicegates powering turbines during both the rising and falling tides.

TIDAL CURRENT DEPOSITS

TIDAL CURRENT DEPOSITS 387

Describe the activity of tidal currents. What role do they play in sediment transport and deposition? Where do deposits form?

Tidal currents scour narrow inlets, keeping them open. Currents carry fine sediment in suspension to be deposited in bays and estuaries.

What is a *mud flat?* Of what is it built? Where is it formed?

Mud flat: an accumulation of fine sediment (including organic matter) built up to level of low tide in bays and estuaries.

Describe a *salt marsh.* Of what material is it formed? How high is it? What kind of streams are found in the salt marsh?

Salt marsh: peat-covered expanse of sediment at level of high tide. Salt marsh is drained by sinuous tidal streams.

How have humans made use of salt marshes? What effects follow such use? How have areas of salt marsh been affected by urbanization?

Human use of salt marsh includes drainage by dikes to create freshwater land for agriculture. Diked land subsides, is vulnerable to saltwater flooding during storms. Salt marsh is filled in urban areas to extend usable land.

COMMON KINDS OF COASTLINES

COMMON KINDS OF COASTLINES 389

What is the concept of uniqueness of a given coastline in terms of changes of sea level, or other cause?

Uniqueness of a given coastline depends upon configuration of land surface or ocean floor against which shoreline has come to rest as a result of *submergence, emergence,* or by the construction of new land.

Define *submergence?* What change of ocean level, or of crust is involved?

Submergence: inundation or partial drowning of a former land surface by a rise of sea level or a sinking of the crust, or both.

Define *emergence.* What change of ocean level, or of crust is involved?

Emergence: exposure of submarine landforms by a lowering of sea level or a rise of the crust, or both.

How can addition of new land create unique coastlines? Give an example.

Addition of new land by such activities as volcanism, delta growth, or coral reef growth can also create unique coastlines.

Name the seven important types of coastlines. Describe and explain each type.

Important types of coastlines: *ria coast, fiord coast, barrier-island coast, delta coast, volcano coast, coral-reef coast, fault coast.*

Ria coast. What processes previously shaped the landmass?

Ria coast: a deeply embayed coast formed by partial submergence of a landmass previously shaped by fluvial processes.

Fiord coast. What processes previously shaped the landmass?

Fiord coast: a deeply embayed, rugged coast formed by partial submergence of glacial troughs. Numerous deep, narrow bays (fiords).

Barrier-island coast. What are the major features of such a coast?

Barrier-island coast: coast with broad zone of shallow water offshore along which a barrier island is built, shutting off a zone of open water (lagoon) from the ocean.

Delta coast.

Delta coast: coast of a delta constructed of sediment brought to the sea at the mouth of a stream.

Volcano coast.

Volcano coast: coast formed by volcano-building and lava flows partly above and partly below sea level.

Coral-reef coast.

Coral-reef coast: coast built out by accumulations of limestone materials secreted by corals and algae.

Fault coast. Against what feature does the water come to rest?

Fault coast: coast in which shoreline comes to rest against the scarp of an active fault. (Down-dropped block may subside below sea level.)

DEVELOPMENT OF A RIA COAST

What prior history of the landmass determines the form of a ria coastline? How are bays formed? What features result in headlands and islands?

Describe the evolution of a ria coast? What features are formed by erosion? What features result from sediment deposition? What form of shoreline ultimately develops?

What are the environmental influences of ria coasts? What commercial activities have been important in the past?

DEVELOPMENT OF A RIA COAST 390

Form of ria coast will reflect the contour of a system of stream-eroded valleys and intervening divides. Submerged valleys form bays; divides form headlands and outlying islands.

Evolution of ria coast: waves form sea cliffs on islands and headlands; sediment then accumulates as beaches, and sandspits, tying islands to mainland and cutting across mouths of bays; offshore islands finally planed off, leaving a simple shoreline in which beaches carry shoreline from one headland to next.

Environmental influences of ria coasts: excellent harbors promote maritime activities of shipbuilding, ocean commerce, fishing.

BARRIER-ISLAND COASTS

Where do barrier-island coasts typically develop? Explain why. With what major feature of the continental margin are these coasts associated? What are the typical forms of the coast?

Define *barrier island.* Of what is it formed? How is it formed?

Define *lagoon.* Where is it situated? With what deposits is it partially filled?

Define *tidal inlet.* Where are inlets found? What keeps them open? How is a new inlet formed?

Describe the environmental influences of barrier-island coasts. How are they modified for navigational use?

Describe the Gulf Coast of Texas as an example of a barrier-island coast. Where are the major ports situated?

BARRIER-ISLAND COASTS 390

Barrier-island coasts are typical of coastal plains bordering broad continental shelves. Emerged sea floors are gently sloping; shallow water extends far offshore. Typical forms are *barrier island* and *lagoon.*

Barrier island: a long, narrow sand embankment built above high tide level by waves and wind and running parallel with an inner mainland shoreline.

Lagoon: expanse of open water between barrier island and mainland shoreline. Often partially filled with mud flats or tidal marsh.

Tidal inlet: a gap or breach in a barrier island, or in any sandspit or sand barrier, lying between the open ocean and a lagoon or bay. Strong tidal currents keep the inlet open, but it may be closed by sediment carried by littoral drift. New inlets are often breached by severe storms.

Environmental influences of barrier-island coasts: shallow water and narrow inlets make natural harbors few and inadequate. Dredging is required, along with jetties in inlets, to maintain deep-water channels.

Gulf Coast of Texas, an example of a barrier-island coast: long, continuous barrier island has few inlets. Galveston built on barrier island at inlet. Other ports are on mainland banks of river mouths.

Describe the barrier islands of the coast of North Carolina.

Barrier islands of North Carolina coast have cuspate headlands (Cape Hatteras, Cape Lookout) and large, wide lagoon. Inner shoreline is deeply embayed, showing submergence.

What is *overwash?*

Overwash: movement of storm swash entirely across a barrier island or barrier beach to reach the lagoon or salt marsh on the inland side.

DELTA COASTS

DELTA COASTS 393

What is a *delta?* Of what materials is a delta formed?

Delta: a deposit of clay, silt, and sand made by a stream at the point where it flows into a body of standing water (lake or ocean). Sediment drops as flow velocity is checked.

What are *distributaries?*

Distributaries: secondary outlet channels formed by subdivision of main stream channel on the delta.

What factors determine the configuration of a delta shoreline? Name four common shapes (outlines) and give one example of each:

Delta shapes depend upon configuration of coastline and upon wave action. Four common shapes:

Arcuate delta

Arcuate delta: delta with shoreline curved convexly outward from the land. Wave action shapes the outer shoreline, making beaches by littoral drift. Example: Nile delta.

Bird-foot delta

Bird-foot delta: delta with long, projecting distributary fingers extending out into open water. Example: Mississippi delta.

Cuspate delta

Cuspate delta: delta with shoreline sharply pointed toward open water. Main river mouth is at point of cusp; sediment is carried to sides by littoral drift. Example: Tiber delta.

Estuarine delta

Estuarine delta: delta built into the lower part of an estuary, or drowned river mouth. Example: Seine delta.

Describe the environmental influences of deltas. How is agriculture facilitated? What is the importance of deltas as sites of major port cities? Name two examples.

Environmental importance of deltas: fertile flat land is intensively cultivated, often has dense population. Important ports lie on deltas at stream mouths (examples: New Orleans, Calcutta, Shanghai). Delta growth is rapid. Keeping stream mouths navigable is a serious engineering problem.

CORAL-REEF COASTS

CORAL-REEF COASTS 396

Describe a *coral reef.* What organisms build reefs?

Coral reef: a rocklike accumulation of mineral carbonate secreted by corals and algae in shallow water along a marine shoreline.

Describe the conditions of coral reef growth. What is the latitude range of reefs? What water conditions are required? To what level are reefs built?

Conditions of coral reef growth: warm water of tropical and equatorial zones between 30°N and 25°S latitude. Water must be clean, free of suspended sediment, and well aerated by wave action. Corals build to level of high tide, give reef a flat top.

Name three types of coral reefs.

Types of coral reefs: *fringing reefs, barrier reefs, atolls.*

Define *fringing reef.* Where is it formed? Where is it widest?

Fringing reef: reef platforms attached to mainland shore. Built widest at exposed locations with strong surf.

Define *barrier reef.* Where is it situated?

Barrier reef: reef separated from mainland shore by a lagoon. Gaps or passes cut across reef.

Define *atoll.* What form has an atoll? How can islands develop? How are atolls formed?

Atoll: circular or closed-loop reef enclosing an open lagoon with no landmass inside. Coral sand may accumulate to form islands on atoll reef.

What is the *subsidence theory* of atolls?

Subsidence theory: hypothesis advanced by Charles Darwin explaining atolls by subsidence of the oceanic crust and continued upbuilding of coral reefs that were originally fringing reefs on a volcanic island.

Describe the environmental aspects of atolls. How are atolls vulnerable to storm destruction?

Environmental aspects of atolls: lack of silicate minerals as plant nutrients, scarcity of fresh water, vulnerability to typhoons and seismic sea waves.

What plant resource is particularly important on atoll islands?

Coconut palm a vital source of food, fiber, and construction material. Fish an important diet item.

RAISED SHORELINES AND MARINE TERRACES

RAISED SHORELINES AND MARINE TERRACES 397

What is a *raised shoreline?* How is it formed? What landforms are present?

Raised shoreline: an inactive (defunct) shoreline raised above level of wave action by a sudden crustal rise. Marine cliff and abrasion platform become elevated landforms.

What is a *marine terrace?* Can they be found in multiple sets?

Marine terrace: former abrasion platform elevated to become a steplike coastal landform. May occur in multiple sets.

Describe the environmental aspects of marine terraces. How are marine terraces used by humans?

Environmental aspects of marine terraces: On a mountainous coastal region terrace provides a continuous flat belt of land suited to railroads and highways and to agricultural and urban uses.

THE LITTORAL CELL AS A MATERIAL FLOW SYSTEM

THE LITTORAL CELL AS A MATERIAL FLOW SYSTEM 398

Describe the concept of a *littoral cell.*

Littoral cell is a coastal compartment that represents an open material flow system.

What are the main components of the system?

System input is sediment from land, which is transported and stored along the shore, lost as output to deep water.

RISING SEA LEVEL AND COASTAL INUNDATION—A THREAT FROM GLOBAL WARMING 399

Define *eustatic.*

Eustatic: an adjective denoting true rise of global sea level as opposed to changes produced by local crustal motions.

In what two ways can eustatic change occur?

Eustatic change can result from either (1) a change in water volume of the world ocean, or (2) a change in the water-holding capacity of the ocean basins.

How does climate change effect sea level?

Global warming will cause thermal expansion of water in the top layer of the ocean creating a larger volume of water and a higher sea level. Global warming will also cause some glacial ice to melt and flow into the oceans, producing a high sea level. If global warming causes more precipitation; then glaciers may grow, cancelling this effect.

What rise in global sea level has been measured since 1900? What is the current trend?

Since 1900, the total eustatic rise has been about 80 mm (3 in.). The current trend is a rapid rate of 2.4 mm (0.1 in.) per year.

What would be the effect of a sea level rise of 1 m (3 ft)?

A sea level rise of 1 m (3 ft) could subject tidal estuaries, marshes, and coastal cities to catastrophic damage from storm surf and storm surges in concert with high tides. Oceanfront cliffs would retreat, resulting in massive losses of shorefront properties.

What is the possible effect of sea level rise on coral reefs?

Coral reefs would grow quickly to maintain their correct level in the face of rising sea level. This could hasten removal of CO_2 from the air and perhaps slow global warming. If the rise is too rapid for reefs to keep pace, they will die.

CHAPTER 20—SAMPLE OBJECTIVE TEST QUESTIONS

A. MATCHING

1.	swash	a. ________	submergence
2.	littoral drift	b. ________	ebb current
3.	progradation	c. ________	marine terrace
4.	tide	d. ________	groin
5.	salt marsh	e. ________	backwash
6.	fiord coast	f. ________	peat
7.	ria coast	g. ________	lagoon
8.	abrasion platform	h. ________	atoll
9.	barrier island	i. ________	glacial trough
10.	coral reef	j. ________	longshore drifting

B. MULTIPLE CHOICE

1. Littoral drift is

   ________a. caused entirely by swash and backwash.

   ________b. responsible for construction of sandspits.

   ________c. caused entirely by longshore drifting.

   ________d. caused by tidal currents.

2. A barrier-island coast

   ________a. has deep water immediately off shore.

   ________b. consists of a barrier island of volcanic rock.

   ________c. is typical of coastal plains bordering shelves.

   ________d. has excellent natural deep-water port facilities.

3. Atolls are most probably formed by

   ________a. a rise of the earth's crust.

   ________b. sand brought from the mainland.

   ________c. collapse of the center of volcano.

   ________d. coral reef upbuilding during subsidence.

C. COMPLETION

1. Coastal uplift, occurring suddenly, results in a ______________________________ ______________________.

2. Cutting back of a beach by wave action is called ______________________________ ______________________.

3. The upper layer of material in salt marsh is a form of ______________________________ ______________________.

4. The tidal current setting in as tide begins to fall is the ______________________ ______________________ current.

5. A variety of coastline produced by tectonic action is the ______________________ ______________________ coast.

6. A coastline type built by stream deposits is a ______________________ ______________________ coast.

7. A gap in a barrier island is called a ______________________.

8. A surge of water moving up the beach slope is the ______________________.

CHAPTER 21

LANDFORMS MADE BY WIND

PAGE

EROSION BY WIND — 401

What two kinds of work does wind perform?

Wind performs two kinds of erosional work: (1) *deflation*; (2) *abrasion*.

Define *deflation*. What sizes of particles are affected by deflation?

Deflation: lifting and transport of loose particles of soil or regolith from dry ground surfaces. Clay and silt sizes are swept up in turbulent winds, carried long distances.

How does *wind abrasion* act? What tools are used by the wind? What minor features are produced by abrasion?

Wind abrasion: sandblast action by fast-moving sand grains impacting rock surfaces; it wars pits, grooves, and hollows in hard rocks.

Explain how wind acts as a sorting agent. What landform is produced by deflation?

Wind acts as a sorting agent, removing fine particles and leaving coarser particles behind. Deflation can produce blowout depressions.

Define *blowout*. How deep and how wide may a blowout be? Where do blowouts form? Under what climate? Do animals contribute to deepening of blowouts? Explain.

Blowout: a shallow depression produced by continued deflation. May be a few meters deep, up to 1 km wide; forms on plains in arid climates. In semiarid grasslands, blowouts form where vegetation is broken by grazing and trampling.

What is the role of deflation on playas of deserts?

On broad playas of desert basins deflation can remove fine-textured sediments over large areas.

What is a *yardang?*

Yardang: long narrow ridge of silty alluvium or old dune deposit shaped by wind abrasion into a streamlined form with prow and tail.

What is a *desert pavement?*

Desert pavement: surface layer of closely fitted pebbles or coarse sand from which the finer particles have been removed by rainbeat and deflation.

DUST STORMS

What is a *dust storm?* What sizes of particles make up a dust storm?

How high into the atmosphere does dust rise in a dust storm? How far can the dust travel?

DUST STORMS 402

Dust storm: dense cloud of silt and fine clay particles suspended in turbulent air mass.

Dust storms rise to heights of several thousand meters, cut visibility to near-zero. Dust may travel as far as 4000 km (2500 mi) before settling back to ground surface.

SAND TRANSPORT BY WIND

What is a *sandstorm?*

What is *saltation?*

SAND TRANSPORT BY WIND 403

Sandstorm: dense, low layer of sand grains traveling in *saltation* over a sand dune or a sand beach. Most grains remain within 2 m (6 ft) of surface.

Saltation: leaping, impacting, and rebounding of spherical sand grains transported over a sand or pebble surface by strong winds. Saltation also causes a slow downwind surface creep of impacted grains.

SAND DUNES

What is a *dune?* What is the distinction between active dunes and in active dunes? What causes dunes to be inactive?

Of what minerals are dune sands composed?

Name the common types of sand dunes.

Describe a *crescent dune (barchan).* What is the slip face? Which way are the points of the crescent directed? Describe the travel of sand grains across the dune surface. Do crescent dunes move? Why?

What is a *slip face?*

SAND DUNES 403

Dune: a hill of loose sand shaped by the wind. *Live dunes* are active dunes bare of vegetation and changing constantly with shifting wind currents. *Fixed dunes* are inactive when covered by plants that hold down the sand grains.

Most dune sand consists of spherical quartz grains. Dunes may also be composed of tephra, shell fragments, gypsum grains, or heavy minerals (e.g., magnetite).

Common types of sand dunes: *crescent dune (barchan), transverse dune (sand sea), parabolic dune (coastal blowout dune, parabolic blowout dune, hairpin dune), longitudinal dune.*

Crescent dune (barchan): hill of loose sand of crescentic base outline, having a sharp crest and steep slip face. Points of crescent are directed downwind. Sand grains travel up gently upwind side of dune, leap over crest and fall on *slip face,* which moves slowly forward. Crescent dunes occur in chains, extending downwind from sand source.

Slip face: the steep face of an active sand dune, receiving sand by saltation over the dune crest and repeatedly sliding because of oversteepening. Angle of repose of slip face is about 35°.

What ar *transverse dunes?* Describe transverse dunes. How is their form related to wind direction? Are slip faces present? What requirements of sand supply must be met?

Transverse dunes: wavelike ridges of free sand, crests trending at right angles to prevailing wind. Entire area resembles storm-tossed ocean surface. Slip faces are present with deep depressions between ridges. Require large sand supply.

What is a *sand sea?*

Sand sea: general term for a large field of transverse dunes.

What are *parabolic dunes?*

Parabolic dunes: isolated sand dunes of parabolic outline, with points directed into the prevailing wind. Typically, a deflation hollow lies within the parabolic dune ridge.

What is a *coastal blowout dune?* Where does this dune occur? Describe the dune form. What is the relation of this dune type to adjacent forest? Give an example of a belt of coastal blowout dunes.

Coastal blowout dune: dune formed adjacent to sand beach of shoreline as deflation produces blowout hollow and heaps sand into horseshoe ridge on landward side. May have slip face, overriding forest. Example: dunes of Indiana Dunes State Park.

What is a *parabolic blowout dune?* Describe the dune form. In what climate does this dune develop? Does the dune move forward? Is there a slip face? Into what elongated dune form can the parabolic blowout dune develop?

Parabolic blowout dune: a low sand ridge of parabolic outline formed to the lee of a shallow blowout in semiarid climate. Sparse plant cover immobilizes the dune ridge. Dunes lack a slip face. Dunes may become elongated into *hairpin dunes* with long, parallel side ridges.

What are *longitudinal dunes?*

Longitudinal dunes: class of sand dunes in which the dune ridges are oriented parallel with the direction of the prevailing wind.

What three main factors influence dune form?

Three main factors that influence dune form are (1) supply of sand; (2) density of vegetation cover (sparse or dense); (3) strength of wind (weak or strong).

DUNES OF THE SAHARA AND ARABIAN DESERT

DUNES OF THE SAHARA AND ARABIAN DESERT 407

Describe some of the unique landforms of the Sahara Desert.

Sahara Desert landforms show structural features in strong relief, shallow basins of alluvial and salt accumulations. Dunes form in extensive fields (*erg*); deflation produces extensive residual surfaces (*reg*).

What is an *erg?*

Erg: large expanse of active sand dunes in the Sahara Desert of North Africa.

What is a *reg?*

Reg: desert surface armored with a pebble layer resulting from long-continued deflation.

What is a *star dune?*

Star dune: large, isolated sand dune with radial ridges culminating in a peaked summit; found in deserts of North Africa and Arabia.

Where are human settlements with agriculture located in the Sahara Desert?

Human habitations are concentrated in a few, widely separated oases where ground water lies near surface.

COASTAL DUNES

COASTAL DUNES 408

What is the common relationship of dunes to the coastline and beaches?

Most low coastlines with beaches in humid regions are bordered by *foredunes.*

What are *foredunes?* What is the form of foredunes? What relationship do plants bear to the foredunes?

Foredunes: a ridge of irregular sand hills typically found adjacent to beaches on low-lying coastlines and bearing a partial cover of plants.

What are *phytogenic dunes?*

Phytogenic dunes: class of sand dunes formed under partial cover of plants; includes foredunes.

Describe the accumulation of foredunes into a barrier ridge. How high is this barrier? What role does the barrier play in coastal processes? What zone does it protect?

Foredunes accumulate to make barrier ridges as high as 90 m (300 ft) above sea level. The ridge serves as a barrier to swash of storm waves, protecting low-lying salt marsh and lagoon on landward side.

Under what circumstances can the foredune ridge be breached? What happens then? In what way can the effects be harmful?

Breaching of foredune ridge may be induced by breakdown of plant cover, allowing blowout hollow to be formed in barrier. Storm overwash may follow seriously damaging ecosystems and properties on landward side.

How does human activity induce the movement of dunes previously stabilized by plant cover? What are the results?

Human-induced movement of dunes occurs when plant cover is broken down by traffic or grazing. Slip faces form; dunes advance over roads, building. Example: Provincelands of Cape Cod.

LOESS

LOESS 409

What is *loess?* Of what material is loess composed? Where is it found? How is it transported?

Loess: deposit of yellowish to buff-colored, fine-grained sediment, largely of silt grade, upon upland surfaces after transport in suspension by wind.

What is *cleavage* in loess?

Cleavage: natural vertical partings in loess, produced by volume shrinkage during compaction. Cleavage tends to cause loess to stand in perpendicular walls and cliffs when undermined.

Where do thick deposits occur? How thick?

Loess deposits are very thick in northern China, up to 90 m (300 ft). Dust source was interior desert Asia.

Name four continental locations where thick loess occurs.

Important loess deposits in United States, central Europe, central Asia, Argentina.

Describe the loess deposits of the Missouri-Mississippi valley region. Where is the loess thickest? From what source was this loess derived?

Loess of Missouri-Mississippi valley region is continuous over prairie plains, ranging from 1 to 30 m (3 to 100 ft). Thickest on east bluffs of river floodplains. Loess derived during late Pleistocene from floodplains of rivers draining wasting ice sheet.

What is the agricultural importance of loess? What types of soils are formed on loess?

Agricultural importance of loess: forms parent matter of Mollisols. Important corn and wheat producing regions on loess soils.

How have humans made use of loess as a structural material?

Loess forms stable slopes, has been excavated for cave dwellings in China and central Europe.

DEFLATION INDUCED BY HUMAN ACTIVITY—THE DUST BOWL 411

What is the concept of plant-cover balance in the semiarid climate, in relation to human activities?

Semiarid climate zones are in delicate state of balance of plant cover; the balance is easily disturbed by human activities.

Describe the Dust Bowl conditions of the 1930s. Where did this activity occur? What agricultural practices may have induced deflation and drifting?

Great Plains region of U. S. experienced severe deflation and dust storms in 1930s, following expanded wheat cultivation. Dust Bowl impoverished many wheat farmers. Topsoil was removed from fields, accumulated in drifts along fences and buildings.

Where is the problem area associated with the dust bowl?

The problem area is on the easternmost border of the semiarid, dry midlatitude climate (steeppe) near the 89th meridian.

Why is this region a problem?

This regions is a problem because average rainfall is insufficient to support sustained agriculture without irrigation, and so devastating droughts will reoccur.

What is being done?

Farmers in the dust bowl region are relying on pumping of ground water from deep aquifers. Ultimately, this water source will decline sharply.

CHAPTER 21—SAMPLE OBJECTIVE TEST QUESTIONS

A. MATCHING

1.	blowout	a. _______	coast
2.	desert pavement	b. _______	Missouri Valley
3.	dune sand	c. _______	slip face
4.	dust storm	d. _______	reg
5.	crescent dune	e. _______	Great Plains
6.	transverse dunes	f. _______	clay
7.	parabolic dune	g. _______	quartz
8.	foredunes	h. _______	hairpin dune
9.	loess	i. _______	sand sea
10.	Dust Bowl	j. _______	deflation

B. MULTIPLE CHOICE

1. Sand grains travel over a dune sand surface by

   ________a. sliding over other grains.

   ________b. rolling over other grains.

   ________c. leaping and rebounding repeatedly.

   ________d. one grain pushing the next.

2. The crescent dune (barchan) has

   ________a. horns pointing upwind.

   ________b. a slip face on the lee side.

   ________c. a star-shaped outline.

   ________d. a dense plant cover.

3. Loess accumulated in greatest thickness

   ________a. in the beds of glacial streams.

   ________b. on east bluffs of river floodplains.

   ________c. on west bluffs of river floodplains.

   ________d. along fence lines in the Dust Bowl.

C. COMPLETION

1. Another name for a large expansive of transverse dunes is ________________________________ ________________________.

2. A dune of parabolic outline formed adjacent to a sand beach is called a coastal ____________ ________________________ dune.

3. The ridge of irregular sand hills typically found adjacent to beaches consists of a dune type called ______________________________ .

4. A layer of wind-deposited material on which rich agricultural soils has formed is made up of ______________________________ .

5. The Great Plains in the 1930s became known as the ______________________________ ____________________ because of wind action.

CHAPTER 22

GLACIAL LANDFORMS AND THE ICE AGE

PAGE

GLACIERS 413

At what thickness does basal ice become plastic, allowing glacier flowage to take place?

Ice sufficiently thick (100 m, 300 ft) becomes plastic at base, moves by slow flowage, becomes a *glacier*.

Define *glacier*.

Glacier: a large, natural accumulation of land ice affected by present or past motion.

What concept relates accumulation of glacial ice to ablation?

Accumulation of glacial ice requires that, on the average, winter snowfall exceeds summer *ablation*.

What is *ablation?*

Ablation: wastage of glacial ice by both melting and evaporation.

What are the stages in transformation of snow to glacial ice?

Glacial ice is formed by compaction of snow into granular ice, then compressed into dense glacial ice.

What are the requirements of altitude or latitude for glaciers to form?

High altitude or high latitude is required for sufficiently low average temperatures for glaciers to form.

What governs the form of alpine glaciers?

Glaciers forming in high mountains occupy preexisting stream valleys, take long narrow shape as *alpine glaciers*.

Define *alpine glacier*.

Alpine glacier: long, narrow mountain glacier on steep downgrade, occupying floor of troughlike valley.

In what way do uplands cause the growth of thick ice sheets?

Uplands in arctic and polar latitudes may intercept enough snowfall by orographic effect to build thick *icecaps*.

Define *icecap*.

Icecap: platelike mass of glacial ice limited to the high summit region of a mountain range or plateau; a variety of glacier.

How can an icecap become a larger type of glacier?

In a glacial period, an icecap can expand greatly, spreading over adjacent lowlands, to form a *continental glacier*, or, *ice sheet*.

What is a *continental glacier (ice sheet)?*

Continental glacier (ice sheet): a large, thick plate of glacial ice moving outward in all directions from a central region of accumulation.

ALPINE GLACIERS

ALPINE GLACIERS 413

Name and describe the important features of alpine glaciers, from upper end to lower end.

Features of alpine glaciers include the cirque, holding the *firn field* in the *zone of accumulation*, *rock steps* with deep *crevasses* and *ice falls*, and a *zone of ablation* in which the ice is heavily charged with rock debris at the *glacier terminus*.

Describe a *cirque.*

Cirque: a bowl-shaped depression holding the collecting ground, or firn field, or an alpine glacier.

What is the *zone of accumulation?*

Zone of accumulation: upper potion of a glacier in which the firn becomes transformed into glacial ice; the zone of nourishment of a glacier.

What is *firn?*

Firn: layers of snow in process of compaction and recrystallization (also called névé).

What is the *firn field?*

Firn field: the expanse of firn in the zone of accumulation.

Describe a *rock step.*

Rock step: an abrupt steepening of gradient of the rock floor beneath an alpine glacier.

What are *crevasses?*

Crevasses: gaping fractures in glacial ice, caused by pulling apart of the upper brittle ice layer.

What is an *ice fall?*

Ice fall: a steepened, deeply crevassed portion of a valley glacier, where it flows over a rock step.

What is the *zone of ablation?*

Zone of ablation: lower portion of a glacier, where ablation exceeds gain of mass by snowfall; the zone of wastage of the glacier.

What is the *glacier terminus?*

Glacier terminus: the lower end, or snout, of a glacier.

Where is glacier motion fastest? Give rough figures for the rate of flow of alpine glaciers.

Glacier motion is fastest near center line and surface, decreasing in speed toward banks and bed. Rate of ice flow is few cm to few m per day in alpine glaciers.

What is the concept of glacier equilibrium? What factors are balanced when equilibrium exists? How is equilibrium upset?

A glacier readily establishes a state of dynamic equilibrium in which rate of accumulation at upper end balances rate of ablation at lower end. Equilibrium is upset by changes in input and output rates, reflecting climatic fluctuations.

What are glacier *surges?*

Surges: episodes of very rapid downvalley movement of an alpine glacier. Surging glacier may develop a sinuous pattern of movement. Movement may be several km in a few months.

GLACIER AS A FLOW SYSTEM OF MATTER AND ENERGY

GLACIER AS A FLOW SYSTEM OF MATTER AND ENERGY 416

Describe a glacier as an open material flow system.

Glacier as an open material flow system has input through solid precipitation, storage and transport in solid state, output by evaporation (sublimation) in vapor state and by melting (liquid state).

What conditions apply in steady state?

In steady state (glacier equilibrium), rate on input (P) balances rate of output (E + R), while storage (G) remains constant.

Describe a glacier as an open energy system.

As an open energy system, a glacier consists of two subsystems: (1) a gravity flow subsystem with conversion of potential to kinetic energy; (2) a thermal subsystem involving input of heat and output as sensible heat and latent heat.

GLACIAL EROSION

GLACIAL EROSION 418

What load does a glacier carry? From what sources is this load derived?

Glacial ice is charged with load ranging from pulverized rock flour to large boulders. Load is derived from rock floor and by falling from rock walls above glacier.

In what two ways does glacial erosion occur?

Glacial erosion occurs by (1) *abrasion*; (2) *plucking*.

Define *glacial abrasion*. What are the evidences of abrasion?

Glacial abrasion: erosion of bedrock by scraping and grinding action of rock fragments held in moving ice.

Define *plucking*. Explain how plucking takes place.

Plucking: lifting out of joint blocks of bedrock as ice surrounds block and then moves forward.

What happens to the debris carried by an alpine glacier?

All rock debris must be left at glacier terminus during ablation; may be carried farther by running water transport.

LANDFORMS MADE BY ALPINE GLACIERS

LANDFORMS MADE BY ALPINE GLACIERS 418

Describe an unglaciated mountain mass, such as that on which alpine glaciers are developed.

Model development visualizes previously unglaciated mountain mass sculptured by weathering, mass wasting, and streams. Profiles are rounded, soils and regolith are thick.

Describe the progressive changes of landscape as alpine glaciers grow in size and integrate into glacier systems. Name the landforms produced by erosion.

Glaciers fill valleys, integrate into systems of tributaries and trunk glacier. Cirques deepen, steep headwalls recede, intersecting in sharp, jagged *arêtes*, with *cols*, and toothlike *horns*.

What is an *arête*?

Arête: sharp, knifelike divide or crest formed between two cirques as the cirque walls retreat on two sides of a mountain.

What is a *horn?* Give an example.

Horn: toothlike peak formed by intersection of three cirque headwalls. Example: Matterhorn.

What is a *col?*

Col: natural pass or low notch in an arête between opposed cirques.

Explain how a glacier forms a U-shaped trough. How do tributary troughs stand in relation to the main trough? What kind of lakes are present after the ice is gone?

Glacier abrasion widens and deepens channel to form a glacial trough of U-shaped cross section. Tributary trough floors lie high above floor of main trough, making *hanging troughs* and *truncated spurs.* Rock basins in higher parts of troughs become lakes, called *tarns*, after ice is gone.

Define *glacial trough.*

Glacial trough: a deep, steep-sided rock trench of U-shaped cross section, formed by alpine glacier erosion.

Define *hanging trough.*

Hanging trough: tributary trough with floor higher than floor of main trough, making an abrupt step down.

What are *truncated spurs?*

Truncated spurs: valley spurs that have been beveled off by glacial abrasion to become part of the sidewall of the main glacial trough.

What is a *rock basin?*

Rock basin: an overdeepened section of the bedrock floor of a cirque or glacial trough forming a depression, often holding a tarn.

What is a *tarn?*

Tarn: lake held in a rock basin.

What are *trough lakes?*

Trough lakes: lakes occupying floors of glacial troughs.

What are *finger lakes?*

Finger lakes: same as trough lakes, but may also refer to lakes occupying basins deepened by a continental ice sheet.

What forms of moraines are associated with alpine glaciers?

Moraines associated with valley glaciers are *lateral, medial, terminal* (*end*), and *recessional* types. Moraine is any accumulation of till.

What is a *lateral moraine?*

Lateral moraine: moraine forming an embankment between the ice of an alpine glacier and the adjacent valley wall.

What is a *medial moraine?*

Medial moraine: moraine riding on the surface of an alpine glacier and composed of debris carried down-valley from the point of junction of two ice streams.

What is a *terminal (end) moraine?*

Terminal moraine (end moraine): moraine deposited as an embankment at the terminus of an alpine glacier or the leading edge of an ice sheet.

What is a *recessional moraine?*

Recessional moraine: moraine produced at the ice margin during a temporary halt in the recessional phase of an alpine glacier or an ice sheet.

GLACIAL TROUGHS AND FIORDS

GLACIAL TROUGHS AND FIORDS 420

What is a *valley train?*

Valley train: a deposit of alluvium extending downvalley from a melting glacier. Valley train may fill the floor of a glacial trough.

What features result from submergence of troughs by the sea?

Close to coastline, deep troughs are partially submerged, becoming *fiords*.

Define *fiord*. Give an example of a region of fiords. In what latitude zone are they well developed? Explain.

Fiord: a narrow, deep, ocean embayment partially filling a glacial trough. Fiords open up as tidewater glaciers melt back. Example: Alaskan fiords. Fiords are abundant on mountainous west coasts of continents in latitudes 50° to 70° N and S, receive heavy precipitation in marine west-coast climate.

ENVIRONMENTAL ASPECTS OF ALPINE GLACIATION

ENVIRONMENTAL ASPECTS OF ALPINE GLACIATION 421

Describe the environmental significance of regions of alpine glaciation. What uses have these regions?

Alpine glaciation terrain sparsely populated, formed barriers between settlements and nations. Striking scenery of wilderness alpine areas a valuable natural resource. Recent increase in use of these areas for winter sports, summer recreation.

How do glacial troughs affect the accessibility of alpine mountains? Give an example.

Deep glacial troughs offer easy access by road and rail lines far into heart of alpine ranges. Example: Adige trough, Italy. Principal passes of Alps are near heads of troughs.

ICE SHEETS OF THE PRESENT

ICE SHEETS OF THE PRESENT 421

Name the two great ice sheets existing today.

Two great ice sheets are (1) Greenland Ice Sheet; (2) Antarctic Ice Sheet.

Describe the Greenland Ice Sheet. Give area of extent, ice thickness, and relationship to landmass. What is the form of the ice surface? What are outlet glaciers? What purpose do they serve?

Greenland Ice Sheet covers 1,740,000 sq km (670,000 sq mi), is 3000 m thick. Only narrow coastal zone of Greenland is icefree. Surface has dome shape. Outlet glaciers reach to sea, forming icebergs.

What are *outlet glaciers?*

Outlet glaciers: ice tongues leading from the ice sheet to the sea through narrow troughs in coastal mountains.

Describe the Antarctic Ice Sheet. Give area of extent and ice thickness. What is the configuration of the rock floor?

Antarctic Ice Sheet covers 13 million sq km (5 million sq mi), is centered on south pole. Ice thickness in places exceeds 4000 m (13,000 ft). Rock floor locally lies 1000–2000 m below sea level. Ice shelves are floating extensions of ice sheet.

What are *ice shelves?* How are ice shelves sustained in growth?

Ice shelves: thick plates of floating ice attached to an ice sheet and fed by the ice sheet and by snow accumulation. Example: Ross Ice Shelf.

SEA ICE — 422

What is *sea ice?*

Sea ice: floating ice of the oceans, formed by direct freezing of ocean water.

At what water temperature does sea ice begin to form? How thick does it get?

Sea ice begins to form at a surface water temperature of about –2°C ($28\frac{1}{2}$°F). Maximum thickness is about 5 m (15 ft).

What is *pack ice?* What are *ice floes* and *leads?*

Pack ice: sea ice that completely covers the sea surface. Individual patches of pack ice are *ice floes;* the narrow strips of open water between floes are *leads.* Under driving force of winds, edges of ice floes are forced upward into pressure ridges.

Compare sea ice conditions of the Arctic Ocean with those of the Antarctic region.

Arctic Ocean retains pack ice in center throughout entire year; partially melts away in summer near shores of surrounding landmasses. Sea ice of Antarctica can drift freely into warmer waters, is limited to about lat. 60°S.

ICEBERGS AND ICE ISLANDS — 424

What are *icebergs?* How do they form? Is the ice fresh or salt?

Iceberg: mass of glacial ice floating in the ocean, derived by *calving* (breaking off) from a glacier that extends into tidewater. Ice is formed of freshwater.

About what proportion of an iceberg lies below the sea surface?

About five-sixths of the iceberg volume lies below water surface.

From what source are icebergs of the North Atlantic derived?

Icebergs of North Atlantic Ocean come from outlet glaciers on Greenland.

What is noteworthy about the form of antarctic icebergs?

Antarctic icebergs are tabular (flat tops, clifflike sides), most are fragments of ice shelves.

What is an *ice island?*

Ice island: large, thick body of floating freshwater ice, derived from a mass of landfast ice. They are found in Arctic Ocean: come from an ice shelf on Ellesmere Island.

ICE SHEETS AND GLOBAL WARMING — 425

What effect would melting of the entire Antarctic Ice Sheet have on sea level?

The melting of the entire Antarctic Ice Sheet would raise global sea level by about 60 m (200 ft).

Describe the theory by which a rapid surge of Antarctic ice raises sea level. How does this occur?

In an area west of the Transarctic Mountains, the normal flow of ice to the ocean is being held back by floating ice sheets that are anchored by being attached (grounded) to the bottom. If the floating ice is thinned at the bottom by melting, it may break loose, allowing the ice on land to surge forward rapidly.

What would be the magnitude of the effect?

The surge of ice flowing into the ocean could raise sea level by 5 m within a short time.

How is global warming predicted to influence the growth of ice sheets?

Global warming may increase precipitation over ice sheets, thus causing them to thicken and build.

Where have effects already been felt?

This effect has already been noted for the Greenland Ice Sheet, which has increased in thickness by 23 cm (9 in.) in the past decade.

What is the probable net effect on sea level?

Enhanced accumulation would slow the rise in sea level produced by melting, perhaps to rates of sea level rise about the same at that experienced in the last decade.

THE ICE AGE

THE ICE AGE 425

What is a *glaciation?*

Glaciation: period of growth and outward spreading of great ice sheets.

What general climatic changes are associated with a glaciation?

Glaciation is associated with general cooling of global air temperatures, but precipitation must remain ample.

What is a *deglaciation?*

Deglaciation: period of rapid shrinkage and partial or complete disappearance of ice sheets, associated with onset of mild climate.

What is *interglaciation?*

Interglaciation: a period in which mild climate prevails following a deglaciation and preceding the next glaciation.

What is an *ice age?*

Ice age: succession of alternating glaciations and interglaciations spanning several millions of years.

What is *The Ice Age?*

The Ice Age: traditional name for ice age prevailing for the past $2\frac{1}{2}$ to 3 m.y. (It spans the Pleistocene Epoch and some end portion of the Pliocene Epoch.)

What is our present position within The Ice Age?

Presently, we are in an interglaciation (Holocene Epoch); it was preceded by the *Wisconsinan Glaciation,* which its maximum about 18,000 years ago and ended rapidly about 15,000 years ago.

Describe the Wisconsinan ice sheets and their locations.

Wisconsinan ice sheets covered much of North America (Laurentide, Cordilleran Ice Sheets), Europe (Scandinavian Ice Sheet), British Isles, Siberia, southern Andes and Patagonia.

EROSION BY ICE SHEETS

EROSION BY ICE SHEETS 427

Describe the erosional activity of ice sheets. What features give evidence of ice abrasion? How are rock knobs shaped by ice action?

Ice sheets erode heavily, removing soil, regolith, and weak types of bedrock. Abrasion features include *striations, chatter marks,* and *crescentic gouges*. Erosion shapes rock knobs into *roches moutonées*.

What are glacial *striations?*

Striations: scratches made by glacial ice abrasion on bedrock outcrops or on the surfaces of rock fragments carred in the ice.

What are *chatter marks?*

Chatter marks: curved fractures produced by ice pressure on the surface of bedrock as angular rock fragments are dragged over the rock surface. Fracture concavity is toward downstream direction of ice motion.

What are *crescentic gouges?*

Crescentic gouges: form of curved fracture opposite in curvature to chatter marks.

What are *roches moutonées?*

Roches mountonées: knobs of bedrock shaped by glacial ice abrasion. Stoss (upstream) side is smoothly rounded; lee side is blocky from plucking of joint blocks.

What are finger lakes? How were they formed? Give an example.

Finger lakes excavated in some areas resemble glacial troughs. Example: Finger Lakes of New York State.

What landscape features are abundant over glaciated areas of shields in North America and Europe?

Shield areas of North America and Europe have countless rock basins eroded by ice and holding lakes.

DEPOSITS LEFT BY ICE SHEETS

DEPOSITS LEFT BY ICE SHEETS 427

What is *glacial drift?*

Glacial drift: general term for all varieties and forms of rock debris deposited in close association with Pleistocene ice sheets.

What are the two major types of glacial drift?

Two major types of glacial drift: (1) *stratified drift*; (2) *till.*

What is *stratified drift*? Of what particle grades is it composed? How was stratified drift formed?

Stratified drift? layers of clay, silt, sand, or gravel deposited by meltwater streams or in bodies of water adjacent to ice.

What is *till?* How is till formed?

Till: heterogeneous mixture of rock fragments ranging in size from clay to boulders, deposited beneath moving ice or directly from ice melting in place.

How thick, on the average, is glacial drift in the United States? How and where does the thickness vary? Where is drift thickest?

Glacial drift in the U. S. averages 6 m (20 ft) thick in mountainous terrain, to 15 m (50 ft) or more on lowlands. Drift in Iowa and Illinois is 45–60 m (150 to 200 ft) thick; thicker in preglacial stream valleys.

What two forms of *till* are recognized?

Ablational till, left by ice melt, is formed of sand, silt, and boulders. *Lodgment till,* dragged beneath ice, is dense, often rich in clay; show poor drainage; is difficult to cultivate because of boulders (erratics).

What are *erratics?*

Erratics: boulders or cobbles carried by moving ice and found enclosed in glacial till or resting on the ground surface.

Describe the various deposits built in the vicinity of a stagnant ice sheet boundary. (See Figure 22.29, pg. 429.)

Deposits near stagnant ice sheet border: terminal moraine becomes knob-and-kettle. Ice lobes left behind curved moraine ridges, joining in interlobate moraines. Recessional moraines formed during temporary halts in ice recession. Outwash plains built out by streams in front of ice sheet. Eskers formed on floors of tunnels under stagnant ice. Drumlins, shaped of till by moving ice, are revealed after ice melts away. Ground moraine spread over general area occupied by ice, later became till plain.

What is *knob-and-kettle?* Of what material is it formed?

Knob-and-kettle: terrain of numerous small knobs and deep depressions along a moraine belt.

What is an *ice lobe?* What form has an ice lobe? What deposit marks its position?

Ice lobe: a forward-bulging extension of an ice sheet having a curved front. Lobes leave a curved moraine.

Describe an *interlobate moraine.* What orientation has this moraine?

Interlobate moraine: moraine formed between two ice lobes; may run approximately parallel with general regional direction of ice movement; merges with terminal moraine.

What is *glaciofluvial sediment?*

Glaciofluvial sediment: sediment derived from glacial ice, transported from stagnant ice masses by meltwater streams and deposited as alluvial material.

What is an *outwash deposit?*

Outwash deposit: accumulation of layers of sand and gravel deposited near the margin of a stagnant ice sheet or alpine glacier.

What is an *outwash plain?* Of what material is it composed? How is it built? What surface features may it show?

Outwash plain: flat plain built of sand and gravel by aggrading streams issuing from stagnant ice border. May contain depressions where stranded ice blocks later melted away.

Describe an *esker.* How is an esker formed? Where is it formed?

Esker: narrow, often sinuous embankment of coarse gravel and boulders, deposited in bed of meltwater streams enclosed in stagnant ice. Eskers often run distances of many tens of km.

Describe a *drumlin.* What is the shape of a drumlin? How is it formed?

Drumlin: hill of till, oval in basal outline, formed by plastering of till beneath moving debris-ladened ice. Occur in swarms. Long axes parallel direction of ice flow.

What is a *till plain?*

Till plain: undulating surface underlain by ground moraine.

What water bodies are found marginal to the ice sheet? What deposits are built into these water bodies? Describe the same area after the ice has disappeared.

Adjacent to ice front, marginal lakes were formed between ice and rising land slopes. Meltwater streams built deltas into the lakes. Later drainage of lakes left deltas standing as delta kames. Flat lake floors contain clay and silt layers, are poorly drained and marshy.

What is a *marginal glacial lake?*

Marginal glacial lake: lake formed of water temporarily ponded between ice front and rising land slopes.

What is a *glacial delta?* Of what material is it formed?

Glacial delta: delta built into marginal glacial lake. Consists of well-sorted sand and gravel with steeply dipping layers.

What are *glaciolacustrine sediments?*

Glaciolacustrine sediments: sediments that have accumulated on the floor of a marginal glacial lake. Sediments are poorly drained, often occupy marshland.

What are *varves?*

Sediments deposited in temporary glacial lakes are silt and clay, often formed into annual bands as *varves.*

What is a *delta kame?* What form has delta kame?

Delta kame: flat-topped hill of sand representing a glacial delta remaining after drainage of marginal lake.

What are *kame terraces?*

Kame terrace: a kame taking the form of a flat-topped terrace built between a body of stagnant ice and a rising valley wall. Most show pitted surfaced from melting of enclosed ice blocks.

ENVIRONMENTAL AND RESOURCE ASPECTS OF GLACIAL DEPOSITS

ENVIRONMENTAL AND RESOURCE ASPECTS OF GLACIAL DEPOSITS 432

What agricultural value have glacial deposits of various kinds? Which deposits make surfaces difficult to farm? Explain. Which deposits make the best farmland?

Agricultural value of glacial deposits ranges widely from poor, stony soils on moraines in hilly terrain to rich farmlands on till plains. Much poorly drained bog land interspersed with uplands, requires drainage of Histosols.

What types of glacial deposits are subject of mass wasting?

Mass wasting, such as earthflows, common on slopes of water-saturated tills.

What special value has knob-and-kettle in terms of suburban development?

Knob-and-kettle of moraine belts highly desirable for larger suburban estates.

What surface layer is responsible for the high quality of soils on till plains of the Middle West?

Till plains of Middle West U. S.largely blanketed by loess forming parent matter of fertile dark Mollisols.

What economic value has stratified drift? Give examples. In what way does glacial drift form valuable ground water resources?

Stratified drift valuable as source of sand and gravel for concrete, base courses of pavements. Drift holds large ground-water supplies, particularly in deep preglacial valleys.

LATE-CENOZOIC ICE AGE

LATE-CENOZOIC ICE AGE 432

Within what three geologic epochs does the Ice Age fall?

The Ice Age falls within the Pliocene, Pleistocene, and Holocene epochs (last three epochs of the Cenozoic Era).

What is meant by the *Late-Cenozoic Ice Age?*

Late-Cenozoic Ice Age: term synonymous with The Ice Age, as used above.

What older methods were formerly used to unravel the history of The Ice Age?

Older methods of research used to unravel details of The Ice Age were based on glacial and related deposits (such as loess) on the continents.

How many glacial stages are represented by deposits in North America? What is a glacial stage? What is an interglacial stage? Name the glacial and interglacial stages of North America in order from youngest to oldest. About how many years ago did the Wisconsinan stage begin?

Major ice advances and recessions of Pleistocene ice sheets: Four separate periods of ice-sheet growth, called glacial stages, interspersed with periods of ice recession to near-disappearance, called interglacial stages. American sequence:

Wisconsinan glacial (began about 60,000 yrs ago.

Sangamonian interglacial

Illinoian glacial

Yarmouthian interglacial

Kansan glacial

Aftonian interglacial

Nebraskan glacial (300,000 to 600,000 yrs ago)

RADIOCARBON DATING

RADIOCARBON DATING 433

What is the *radiocarbon dating method*?

Modern research on chronology of the Late-Cenozoic Ice Age began with introduction of the *radiocarbon dating method.*

What other method has been used in connection with the radiocarbon method?

Tree-ring records (dendrochronology) have been used to supplement radiocarbon method.

OXYGEN-ISOTOPE CHRONOLOGY OF DEEP-SEA CORES

OXYGEN-ISOTOPE CHRONOLOGY OF DEEP-SEA CORES 433

What research methods on ice-age history have been applied to sediments on the ocean floors?

Method of magnetic polarity reversals, applied to core sediments of deep-sea floor, have revealed earlier Cenozoic record of numerous glaciations.

How can cycles of global climate change be determined from deep-sea cores?

Oxygen-isotope method reveals cycles of colder and warmer surface climate, expressed as a *paleoglaciation curve,* marked off into *isotope-ratio stages,* going back as far as –2 to –3 m.y., or more.

What number of glaciations is revealed by the *paleoglaciation curve*?

Paleoglaciation curve shows as many as 30 glaciations, spaced at intervals of about 90,000 years.

ORIGIN OF THE GREAT LAKES

ORIGIN OF THE GREAT LAKES 434

What landscape existed in the Great Lakes region prior to the Pleistocene glaciation? What effect did the glacial ice have on this area?

Prior to glaciation, Great Lakes region was lowland of weak rock occupied by major streams. Pleistocene glacial ice deeply eroded the weak rock lowland, carried till to the south to form blocking morainal deposits.

When and how did the earliest of the ancestral Great Lakes come into existence?

As ice receded, earliest marginal glacial lakes were formed (Lake Chicago, Lake Maumee).

By what route did the early marginal lakes drain?

Drainage of early marginal lakes was south into Mississippi River system; then an eastward drainage was established into Mohawk Valley.

What drainage line was used following total disappearance of the ice?

Following final ice disappearance, drainage was established through Ottowa River and St. Lawrence River. Crustal tilting led to abandonment of Ottawa River outlet.

ALPINE AND PERIGLACIAL ICE-AGE ENVIRONMENTS

ALPINE AND PERIGLACIAL ICE-AGE ENVIRONMENTS 436

What scientific evidence is there of lowered global air temperatures during glaciation?

Lowered elevation of the snowline over a wide range of latitude is evidence of lowered air temperature during a glaciation.

What is the *snowline?*

Snowline: lower elevation limit of the zone of snow accumulations (snowbanks) that last through the entire year.

What amount of lowering of air temperature is indicated?

Drop in mean annual global air temperature of between 5 and 7 C° (9 and 13 F°) is indicated by lowering of snowline.

What is the *periglacial* environment?

Periglacial: pertaining to the physical (climatic) environment near the border of an ice sheet. Climate is frost-controlled tundra with features of ground ice and intense freeze-thaw activity.

What periglacial features remain today in middle latitudes?

Near former ice borders relict features of ground ice are found today in soil and regolith in middle latitudes.

What other climatic shifts took place during glaciation?

During glaciation atmospheric circulation patterns were shifted; upper-air westerlies moved to lower latitudes, as did the subtropical high-pressure belt and its deserts. Wet equatorial belt was narrowed and had lowered precipitation.

PLUVIAL LAKES OF THE GREAT BASIN

What effect did glaciation have on inland lakes of the western United States?

What are *pluvial lakes?*

PLUVIAL LAKES OF THE GREAT BASIN 436

During glaciations, lakes of closed basins of the Basin and Range region expanded because of increase of runoff input and lowered evaporation losses. Many new lakes were formed.

Pluvial lakes: lakes that formed, or expanded greatly, during glaciations. About 120 such lakes existed during Wisconsinan maximum. Example: Lake Bonneville in location of present-day Great Salt Lake, Utah.

ICE-AGE CHANGES OF SEA LEVEL

How was sea level affected by glaciations?

How much was the drop in sea level during the Wisconsinan glaciation?

What effect did sea level lowering have on continental shelf?

How did sea level changes affect the action of streams and rivers?

ICE-AGE CHANGES OF SEA LEVEL 437

During glaciations ocean water was withdrawn to form ice sheets, causing lowering of sea level.

Maximum drop in sea level was –60 to –80 m (–200 to –260 ft) at about –18,000 years (Wisconsinan maximum).

Continental shelf of eastern North America was exposed widely, with shoreline moved 100 to 200 km (60 to 125 mi) east of its present position.

During lowered sea level, streams were extended across the widened shelf and trenched (degraded) their valleys; during high sea level streams aggraded; result was formation of nested alluvial terraces.

ICE SHEETS AND CRUSTAL REBOUND

What effects did the growth of ice sheets have on crust beneath?

What landforms have been produced by crustal rebound?

ICE SHEETS AND CRUSTAL REBOUND 438

Ice sheets caused crust to be depressed; when ice sheets disappeared crustal rise, called *postglacial rebound,* occurred.

Crustal rebound has resulted in emergence of ancient beaches as elevated shorelines.

CAUSES OF ICE AGES

On what two levels can the problem of causes of ice ages be treated?

CAUSES OF ICE AGES 439

Causes of multiple glaciations (ice ages) can be treated on two levels: (1) basic (fundamental) conditions favoring or causing an ice age; (2) immediate or forcing causes precipitation a glaciation or a deglaciation.

FUNDAMENTAL CAUSES OF AN ICE AGE

Name four fundamental factors that might tend to cause an entire ice age.

FUNDAMENTAL CAUSES OF AN ICE AGE 439

Four fundamental factors that might tend to cause an entire ice age include: (1) favorable position of continents with respect to poles; (2) withdrawal of shallow oceans from continents; (3) sustained volcanic activity; (4) sustained period of diminished intensity of solar energy reaching earth.

Comment on (1) the favorable position of continents with respect to poles, and (2) withdrawal of shallow oceans from continents.

Commentary on fundamental factors:

(1) Following breakup of Pangaea and opening of Atlantic Ocean basin, North America and Greenland moved into high latitudes, enclosing the Arctic Ocean. This favored growth of Cenozoic ice sheets on northern continents, aided by blockage of flow of warm ocean currents into Arctic Ocean.

(2) Near close of Mesozoic Era (late Cretaceous time) shallow continental seas were extensive. By mid-Miocene time, seas had retreated to edges of continents. This would perhaps tend to intensify cold of continental winters.

VOLCANIC ACTIVITY AND ICE AGES

Comment on (3) sustained volcanic activity, and (4) sustained period of diminished intensity of solar energy reaching the earth.

VOLCANIC ACTIVITY AND ICE AGES 440

(3) Large quantities of fine volcanic dust in upper atmosphere would cause blockage of solar energy reaching earth's surface, causing climate cooling. Evidence is in dispute.

(4) Sun's energy output may increase or decrease during an ice age, affecting earth climate. Evidence is lacking of what may have happened.

CAUSES OF GLACIATIONS AND INTERGLACIATIONS

Briefly state four hypotheses of triggering of glaciations.

CAUSES OF GLACIATIONS AND INTERGLACIATIONS 441

Four hypotheses of immediate causes of cyclic glaciations: (1) triggering by bursts of volcanic activity; (2) control by astronomical cycles (astronomical hypothesis); (3) control by changes in arctic sea-ice cover; (4) changes in albedo of land and water surfaces.

Describe the *astronomical hypothesis.*

Astronomical hypothesis: known cyclic variations in geometry of earth's orbit and tilt of axis have generated periods of reduced insolation sufficiently intense to cause glaciations.

What two factors are involved in the astronomical hypothesis?

Two factors involved in the astronomical hypothesis include: (1) the changing distance between earth and sun (21,000–yr cycle in shift of perihelion through the year); and (2) the changing angle of tilt of the earth's axis of rotation (41,000–yr cycle of change in tilt of earth's axis). Two factors are combined, yielding strong peaks and valleys in insolation at latitude 65°N.

What is the *Milankovitch curve?* What cycles are strongest? (See Figure 22.45, pg. 441.)

Milankovitch curve: graph showing the variations in insolation at a given earth latitude, such as 65°N. High insolation peaks at average intervals of 80,000 to 91,000 years.

Does other evidence favor the astronomical hypothesis?

Isotope-ratio cycles of deep-sea cores correlate with Milankovitch cycles.

Do oxygen-isotope rations in ice-sheet cores support the astronomical hypothesis? What does the evidence show?

Yes, oxygen-isotope data support the astronomical hypothesis. The evidence shows ice volume increasing slowly through each major cycle, then falling dramatically at its close.

Describe the "conveyor belt" of ocean circulation and its effect on ice volume.

The "conveyor belt" is a flow of warm surface water from the tropics to the subarctic zone in the eastern North Atlantic Ocean. This warm water releases and enormous amount of heat, retarding the growth of ice sheets. The water is soon chilled and sinks to great depths, slowly moving through the Indian and Pacific Oceans to emerge off the western coast of North America.

What happens to this circulation pattern during glaciation? How may this occur?

During glaciation, the pattern is disrupted, so that its warming effect is not in operation. The change may be produced by wind shifts, produced in turn by changes in Hadley cell circulation and prevailing westerlies.

HOLOCENE CLIMATE CYCLES

HOLOCENE CLIMATE CYCLES 442

What is the Holocene Epoch?

Holocene Epoch: last epoch of geologic time, commencing about 10,000 years ago; it followed the Pleistocene Epoch.

What climatic change marked the onset of the Holocene Epoch?

At onset of Holocene, ocean surfaces warmed rapidly; continental climate zones shifted rapidly poleward; plants became reestablished in previously glaciated areas.

What was the *Boreal stage?*

Boreal stage: the first climatic stage of the Holocene Epoch. Plants of the midlatitude zone were similar to those of the present boreal forest zone.

What was the *Atlantic climatic stage?*

Atlantic climatic stage: warmer-than-average climatic stage from about –8000 to –5000 yr; it was a *climatic optimum.*

What is a *climatic optimum?*

Climatic optimum: a past period of climate warmer than the present climate.

What was the *Subboreal climatic stage?*

Subboreal climatic stage: a climatic stage with below-average temperatures, spanning the period about –5000 to –2000 yr. Glaciers readvanced during this stage.

When did a secondary climatic optimum occur?

A secondary climatic optimum occurred about –1000 to –800 yr (A. D. 1000 to 1200).

What was the *Little Ice Age?*

Little Ice Age: climatic episode of below normal temperatures from about –550 to –150 yr (A. D. 1450–1850) during which time alpine glaciers advanced to lower levels.

How does evidence from ice cores relate to Holocene climate changes?

Oxygen-isotope ratios, measured in layers of ice cores of Greenland and Antarctic ice sheets, reveals cycles of air temperature change as far back as 150,000 years (through Wisconsinan Glaciation). Sudden shift to warmer climate at close of Wisconsinan is strongly shown. Core for past 800 years show that cycles of warming and cooling are normal to Holocene.

CHAPTER 22—SAMPLE OBJECTIVE TEST QUESTIONS

A. MATCHING

1.	ablation	a. _______	hill
2.	cirque	b. _______	fiord
3.	tarn	c. _______	ice tunnel
4.	trough	d. _______	marginal lake
5.	ice sheet	e. _______	ice wastage
6.	moraine	f. _______	rock basin
7.	outwash plain	g. _______	stratified drift
8.	esker	h. _______	firn
9.	drumlin	i. _______	ice shelf
10.	delta kame	j. _______	interlobate

B. MULTIPLE CHOICE

1. Glacier equilibrium exists when

   _______a. ablation exceeds accumulation.

   _______b. accumulation exceeds ablation.

   _______c. ablation is equal to accumulation.

   _______d. climate becomes warmer.

2. The basic requirements for continental glaciation to occur include

   _______a. a general lowering of the continents.

   _______b. a general rise in average global temperature.

   _______c. an increase in intensity of solar energy.

   _______d. a lowering of average global temperature.

3. An esker is an embankment

   _______a. formed within a tunnel in stagnant ice.

   _______b. heaped up by a forward moving ice sheet.

   _______c. built into a marginal lake.

   _______d. of till formed between tow ice lobes.

C. COMPLETION

1. Glacial erosion occurs by ______________________ and ______________________.
2. A toothlike summit left by alpine glaciation is a ______________________ ______________________.
3. A glacial trough partially submerged by the ocean is a ______________________ ______________________.

4. The most recent glacial stage in North America is the ______________________________ ______________________.

5. The two major types of deposits left by ice sheets are ______________________________ ______________________ and ______________________________.

6. A moraine formed between two ice lobes is called an ______________________________ ______________________.

CHAPTER 23

THE SOIL LAYER

PAGE

THE DYNAMIC SOIL

THE DYNAMIC SOIL 445

What concept relates to the soil as a dynamic body?

Concept: the soil is a dynamic layer, continually experiencing many complex physical and chemical activities and changes.

In what three forms does the substance of the soil exist?

Substance of soil exists in three states: (1) solid; (2) liquid; (3) gas.

What is *pedology?*

Pedology: soil science; science of the soil as a natural surface layer capable of supporting plants.

THE NATURE OF SOIL

THE NATURE OF SOIL 445

Define *soil.*

Soil: a natural terrestrial surfaced layer containing living matter and supporting or capable of supporting plants.

What kinds of substances are included in the soil?

Soil includes both inorganic (mineral) and organic matter, the latter both living and dead.

Can the lower limit of the soil be defined?

Lower limit of the soil is not easily or simply defined. Nonsoil beneath is devoid of roots or signs of biologic activity.

What is a *soil horizon?*

Soil horizon: a distinctive layer of the soil, more or less horizontal, set apart from other soil zones or layers by differences in physical and chemical composition, organic content, structure, or a combination of those properties, produced by soil-forming processes.

How long a span of time is required to form a soil with distinctive horizons?

Time required to form a soil with horizons is highly variable. Minimum time is on order of 100 to 200 years; equilibrium state may require thousands of years to attain.

CONCEPT OF THE PEDON

CONCEPT OF THE PEDON 446

What is the concept of the *polypedon*?

Polypedon: the smallest distinctive division of the soil of a given area; it has a distinctive set of properties different from properties of adjacent polypedons. The polypedon is visualized as composed of *pedons*.

Explain the concept of the *pedon*.

Pedon: a soil column extending from the surface down to a lower limit in some form on nonsoil, such as regolith or bedrock. Pedon can be visualized as a hexagonal column displaying a *soil profile* on each side.

What is a *soil profile?* Name the major soil horizons of the soil profile.

Soil profile: display of soil horizons on the face of a pedon or on any freshly cut exposure through the soil. Major horizons are A, B, C and E.

Of what does the *soil solum* consist?

The *soil solum* consists of the A, E, and B horizons of soil profile.

What is the nature of the C horizon?

The C horizon is the parent material of the soil solum; it lies below the limit of root activity.

SOIL COLOR

SOIL COLOR 447

In what way does soil color reveal soil composition?

Soil color may reveal soil composition. Black indicates presence of organic matter (humus); red reveals iron sesquioxide (hematite).

How is soil color described in soil science? What three color variables are evaluated?

Soil color description is based on books of standard colors (Munsell color system). Measurable variables: hue, value, chroma.

SOIL-TEXTURE CLASSES

SOIL-TEXTURE CLASSES 447

What are *soil-texture classes*?

Soil-texture classes: classes of the mineral portion of the soil based on varying proportions of sand, silt, and clay, expressed as percentages.

What is *loam?*

Loam: soil-texture class in which no one of the three grades dominates. Example: sand 40%, silt 40%, clay 20%.

What soil properties are related to soil texture?

Soil texture determines soil-water retention and transmission properties.

How is soil texture related to soil-water storage capacity?

Soil texture is directly related to soil-water storage capacity: low values for sandy textures, high for clay-rich textures.

How is soil texture related to drainage properties of the soil?

Sandy textures drain rapidly; clay textures drain slowly.

What is the *wilting point?*

Wilting point: a critical value of soil-water storage less than which plant foliage wilts.

SOIL CONSISTENCE

SOIL CONSISTENCE 448

What is *soil consistence?*

Soil consistence: quality of stickiness of wet soil, plasticity of moist soil, and degree of coherence or hardness of soil when it holds small amounts of moisture or is in the dry state.

What is cementation? What causes cementation to occur?

Cementation: hardening of a soil horizon so that it does not soften when wetted; caused by accumulation of mineral substances ($CaCO_3$ SiO_2, iron oxides).

SOIL STRUCTURE

SOIL STRUCTURE 448

What is *soil structure?*

Soil structure: presence, size, and form of aggregations (lumps or clusters) of soil particles.

What is a *ped?* How are peds described?

Ped: an individual natural soil aggregate. Peds are described in terms of shape, size and durability.

What are the four primary types of soil structure?

Four primary types of soil structure: platy, prismatic, blocky, spheroidal.

What are *cutans?* How are they formed?

Cutans: thin films or coating on soil peds or on individual coarse mineral grains. Clay cutans (clay skins) form when clay particles are carried down through the soil.

SOIL HORIZONS

SOIL HORIZONS 450

What are the two major classes of soil horizons?

Two major classes of soil horizons: *organic horizons, mineral horizons.*

What is an *organic horizon?* How are they designated?

Organic horizon: soil horizon overlying the mineral horizons and formed of accumulated organic matter derived from plants and animals. Designated by letter O.

What are the subdivisions of the organic horizon?

Subdivisions of organic horizon: Uppermost, O_i, consists of vegetative matter recognizable to unaided eye; beneath is O_a, consisting of unaltered remains of plants and animals not recognizable, i.e., *humus.*

What is *humus?*

Humus: dark brown to black organic matter consisting of fragmented plant tissues partly oxidized by consumer organisms.

What is *humification?*

Humification: process by which humus of O_a horizon is produced.

Of what basic groups of substances are mineral soil horizons composed?

Mineral soil horizons are formed of two basic groups of materials: (1) skeletal minerals; (2) clay minerals.

What are *skeletal minerals?*

Skeletal minerals: minerals grains, mostly of sand and silt grades, that make up the chemically inactive fraction of the soil, as distinguished from the clay minerals.

Describe the mineral horizons of the soil. How is each designated?

Mineral horizons: A and B horizons; they have less than 20% organic matter when no clay is present, less than 30% when clay is 50% or more.

Describe the *A horizon*.

A horizon usually rich in organic matter, dark in color.

Describe the *E horizon*.

E horizon, often pale, shows loss of clay minerals and oxides of Fe and Al, high residual concentrations of quartz sand or coarse silt.

Describe the *B horizon*.

B horizon shows gain of mineral matter, brought down from horizons above. May have high concentrations of clay minerals, oxides of Fe and Al, and organic matter (humus); is often dense and tough, may be cemented.

Describe the *C horizon*.

C horizon is weathered parent material (regolith, alluvium) little affected by biologic activity, not part of the solum. May have accumulations of $CaCO_3$, silica, or soluble salts.

Describe the *R horizon*.

R horizon: bedrock underlying the C horizon(or the B horizon when the C horizon is missing).

THE SOIL SOLUTION

THE SOIL SOLUTION 450

What is the *soil solution?*

Soil solution: aqueous solution held in the soil as soil water and containing dissolved atmospheric gases and ions.

Which atmospheric gases are dissolved in the soil solution?

Important dissolved gases: O_2, N_2, CO_2.Carbonic acid formed by solution of CO_2 and organic acids play major role in chemical processes.

IONS

IONS 451

What is an *ion*?

Ion: atom or group of atoms bearing an electrical charge as the result of gain or loss of one or more electrons. Ions may be *cations* or *anions*.

Define *cation*.

Cation: positively charged ion.

Define *anion*.

Anion: negatively charge ion.

List the important cations in the soil solution. Give the chemical symbol for each.

Important cations in the soil solution:

H^{+}	Hydrogen
Al^{+++}	Aluminum
$Al(OH)^{++}$	Hydroxyl aluminum
Ca^{++}	Calcium
Mg^{++}	Magnesium
K^{+}	Potassium
Na^{+}	Sodium
$NH4^{+}$	Ammonium

List the important anions in the soil solution. Give chemical symbol for each.

Important anions in the soil solution:

Cl^{-}	Chlorine
SO_4^{--}	Sulfat
OH^{-}	Hydroxide
HCO_3^{-}	Bicarbonate
NO_3^{-}	Nitrate

What are the atmospheric sources of ions in the soil solution?

Atmospheric sources of ions in soil solution: rainwater containing atmospheric particulates and pollutants, including mineral dusts, sea salts, combustion products, volcanic emissions.

SOIL COLLOIDS AND CATION EXCHANGE

SOIL COLLOIDS AND CATION EXCHANGE 451

Describe colloidal clay particles.

Colloidal clay particles are thin, platelike bodies with great surface area relative to volume.

Describe the crystal structure of clay minerals.

Clay mineral crystal structure consists of *crystal lattices* in which thin parallel *lattice layers* are repeated. Such minerals are *layer silicates.*

What form of chemical activity affects the layer silicates?

Lattice layers are readily penetrated by free ions of metals and water.

Explain how clay particles can hold cations.

Surface of clay particle, composed of oxygen atoms, is negatively charged, can attract and hold cations.

Which of the cations are most commonly held to clay colloids?

Common ions held by clay colloids: H^{+}, Al^{+++}, Na^{+}, K^{+}, Ca^{++}, Mg^{++}.

What is cation exchange?

Cation exchange: replacement of certain cations by other cations on the surfaces of colloidal clay particles, following a replacement order. Al and H ions have highest priority.

What is *cation-exchange capacity?*

Cation-exchange capacity (CEC): capacity of a given soil to hold and to exchange cations, stated in milliequivalents.

What is the *milliequivalent*?

Milliequivalent: ration of weight of ions to weight of soil (incomplete definition).

Which minerals have high CEC? Which have low CEC?

Minerals of high CEC: vermiculite, montmorillonite. Intermediate CEC: illite. Low CEC: kaolinite. Very low CEC: sequioxides of Al and Fe. Humus particles may have very high CEC.

As soils change with the passage of long spans of time, how does clay mineral composition change?

Young soils in early stage of development are rich in minerals of high CEC; as soils evolve, minerals are altered to those of low CEC.

How does CEC affect soil fertility?

Soils of high CEC are potentially of high fertility; those of low CEC are of low fertility.

SOIL ACIDITY AND ALKALINITY

SOIL ACIDITY AND ALKALINITY 452

What are the two general classes of readily exchangeable cations in the soil?

Two general classes of readily exchangeable cations in soils: (1) *base cations (bases;* (2) *acid-generating cations.*

Define *base cations.*

Base cations (bases): certain cations in the soil solution that are also plant nutrients; most important are cations of Ca, Mg, K, Na.

Under what conditions is the soil *alkaline?*

Alkaline soil: soil in which the majority of cations held by soil colloids are base cations.

Define *acid-generating cations.*

Acid-generating cations: cations, mostly of hydrogen and aluminum, whose presence in large numbers in the soil solution produces an *acid* condition (pH lower than 7).

Name the acid-generating cations.

Acid-generating cations:

Al^{+++}	Aluminum
$Al(OH)^{++}$	Hydroxyl aluminum
H^{+}	Hydrogen

In terms of cations present, when is a soil *acid* in reaction?

Acid soil: soil in which total numbers of acid-generating cations comprise from 5 to 60% of the exchangeable cations present.

What is the *soil pH?*

Soil pH: a number denoting the degree of acidity of alkalinity of the soil solution. The pH number expresses the measure of concentration of hydrogen ions. A pH of 7.0 is neutral; below 7 is acid; above 7 is alkaline.

What range of pH is found in soils?

The pH of soils ranges from 4 (very strongly acid) to 10 (strongly alkaline).

Why is an acid condition undesirable for most agriculture? How is acidity corrected?

Acid condition is unfavorable to most food and forage crops, because nutrient base cations are few. Addition of *lime* raises soil pH to weakly acid or neutral level.

What is *lime?*

Lime: calcium oxide (CaO) or calcium carbonate ($CaCO_3$). Finely ground pure limestone, largely $CaCO_3$ is most widely used in agriculture.

Which acid-generating cations are dominant in highly acid soils?

Exchangeable Al^{+++} and AL $(OH)^{++}$ are dominant acid-generating cations in highly acid soils.

About what percent of the cations are exchangeable bases in a neutral soil solution?

In neutral soil solution, pH=7, exchangeable bases form 85% of the total CEC.

Which base cation dominates in highly alkaline soils?

Sodium ion, Na^+ reaches high concentration in highly alkaline soils, making soil toxic to many plant species.

BASE STATUS OF SOILS

BASE STATUS OF SOILS 453

How are soils stratified in a classification based on fertility for food crops?

Soils are assigned status levels, determined by *percentage base saturation.*

Define *percentage base saturation.*

Percentage base saturation (PBS): percentage of exchangeable base cations present with respect to total CEC of the soil.

What are soils of *high base status?*

Soils of *high base status,* PBS greater than 35%; *low base status,* less than 35%.

Why is base status important in agriculture?

Soils of high base status have high level of natural fertility; those of low base status require applications of fertilizers.

SOIL-TEMPERATURE REGIMES

SOIL-TEMPERATURE REGIMES 453

What importance has soil temperature in soil processes?

Biological and chemical processes are greatly slowed or prevented by low soil temperatures, accelerated by warm temperatures. Root growth and seed germination are strongly temperature dependent.

What is a *soil-temperature regime?*

Soil-temperature regime: characteristic annual cycle of soil temperature defined in terms of mean annual soil temperature (T) and the difference between mean temperatures of warm and cold seasons (Ts – Tw).

Name the soil-temperature regimes and give definitions for each. (See Table 23.4, pg. 454.)

Soil-temperature regimes:

Name	T	(Ts-Tw)
Pergelic	$T<0^\circ$	
Cryic	$0^\circ<T<8^\circ$	
Frigid	$T<8^\circ$	$>5^\circ$
Mesic	$8^\circ<T<15^\circ$	$>5^\circ$
Thermic	$15^\circ<T<22^\circ$	$>5^\circ$
Hyperthermic	$T<22^\circ$	$>5^\circ$

What prefixes of soil subclass names are derived from the soil-temperature regimes?

Prefixes of soil subclass names derived from soil-temperature regimes:

Cryic	Cry–, Cryo–
Frigid	Bor–
Thermic	Trop–

SOIL-WATER REGIMES

SOIL-WATER REGIMES 454

How are soil-water regimes used in soil science and soil classification?

Soil-water regimes (see Chapter 9) are adapted for describing soil-water conditions typical of soils of various classes.

What is a dry soil? a moist soil?

Dry soil: soil in which soil-water tension is 15 bars or greater. Moist soil has tension less than 15 bars.

What is a *bar?*

Bar: unit of moisture tension equal to about one atmosphere of barometric pressure (approx. 1000 mb).

Define *soil water control section:*

Soil-water control section: the portion of the soil that is used to define dry and moist conditions. A typical control section might be between depth of 20–60 cm (8–24 in.).

Name the five soil-water regimes used in soil classification. Define each and give its climate affiliation.

Soil water regimes:

Aquic regime: soil is saturated most of the time; bog condition.

Udic regime: soil not dry as long as 90 days per year. Associated with moist climates with little or no soil-water deficiency. (*Perudic* regime is associated with perhumid climate subtypes.)

Ustic regime: soil is dry for 90 or more cumulative days in most years. Associated with semiarid subtypes of dry climates.

Aridic (torric) regime: warm soils never moist in some or all parts as long as 90 consecutive days. Applies to semidesert and desert subtypes of dry climates.

Xeric regime: soil is dry in all parts for 45 days or more in dry summer season; moist in all parts for 45 or more days in rainy winter season. Applies to Mediterranean climate; semiarid, semidesert, and subhumid subtypes.

What prefixes are derived from the names of the soil-water regimes?

Prefixes for soil subclass names derived from soil-water regimes:

Aquic	Aqu–
Udic	Ud–
Ustic	Ust–
Aridic	Aridi–
Torric	Torr–
Xeric	Xer–

LANDFORM AND SOIL

LANDFORM AND SOIL 455

What is meant by *landform?*

Landform: (singular) general term for total configuration of the ground surface as a factor in soil formation; includes *slope, aspect,* and *relief.*

Define *slope.*

Slope: angle of inclination of the ground or soil surface with respect to the horizontal.

Define *aspect.*

Aspect: orientation of an element of slope in terms of compass direction (facing south, facing north, etc).

Define *relief.*

Relief: average difference in elevation between adjacent high and low points, such as hilltops and valley floors.

How does landform influence the soil profile?

On upland surfaces of nearly zero slope, soil profiles and horizons are thicker than normal; on steep slopes, profiles and horizons are thin. Low-lying valley floors and floodplains are poorly drained; here organic horizons are thick. Slope aspect influences soil-temperature and soil-water regimes.

BIOLOGIC PROCESSES IN SOIL FORMATION

BIOLOGIC PROCESSES IN SOIL FORMATION 455

In what two basic ways do biological processes contribute to soil formation?

Biologic processes contribute to soil forming processes in two basic ways: (1) production of organic matter (humus); (2) recycling of nutrients from soil to plant structures and back to soil.

How do animals living in the soil affect soil structure and composition?

Some animals burrow into soil, creating openings; others rework soil in digestive tracts.

How is soil evolution influenced by the natural plant cover?

Changing succession of plant species (Chapter 26) on newly exposed or deposited parent soil material (alluvium, dunes, till) contributes to a succession of changing soil properties and to soil horizon formation.

REVIEW OF BASIC PEDOGENIC PROCESSES

REVIEW OF BASIC PEDOGENIC PROCESSES 456

What are *pedogenic processes?*

Pedogenic processes: group of recognized basic soil-forming processes, mostly involving the gain, loss, translocation, or transformation of materials within the soil body.

What is *soil enrichment?*

Soil enrichment: additions of materials to the soil body. Materials include sediment deposited on soil by wind (loess, volcanic ash) or water (colluvium), and organic matter accumulating in the O horizon.

What is *leaching?*

Leaching: pedogenic process in which material is lost from the soil by downward washing out and removal by percolating surplus soil water.

What two forms of translocation of soil material are related to a moist (udic) soil-water regime?

Translocation of soil materials in udic or perudic soil-water regime involve both *eluviation* and *illuviation.*

Define *eluviation.*

Eluviation: pedogenic process consisting of the downward transport (translocation) of fine particles, particularly colloids (both mineral and organic), carrying them out of an upper (E) horizon.

What horizon feature results from eluviation?

Eluviation may leave behind the coarse skeletal minerals, particularly quartz grains of sand and silt grades; resulting in *silication.*

What is *silication?*

Silication: increase in proportion of silica in a soil horizon experiencing eluviation.

Define *illuviation.*

Illuviation: accumulation in a lower soil horizon (B horizon) or materials brought down from a higher horizon (E horizon).

What materials are deposited in the B horizon by illuviation?

Materials accumulated by illuviation in B horizon may include clay particles, organic particles (humus), sesquioxides of Fe and Al, (also some forms of silica).

What processes are involved in the translocation of carbonate matter?

Translocation of calcium carbonate ($CaCO_3$) involves both *decalcification* and *calcification.*

What is *decalcification?* Under what soil-water regime does it occur?

Decalcification: removal of $CaCO_3$ through carbonic reaction in soils of the udic soil-water regime, or in wet season of a wet-dry climate.

What is *calcification?* Under what soil-water regime does it occur?

Calcification: accumulation of $CaCO_3$ in the B or C horizon of the soil. Associated with the ustic and aridic soil-water regimes in dry climates.

What are *salinization* and *desalinization?* Where do they occur?

Salinization: precipitation of soluble salts within the soil, through evaporation of soil water in a dry climate or dry season (ustic, aridic, xeric regimes).

Desalinization: removal of soluble salts from the soil in solution by action of surplus soil water received in rainy season or by irrigation in a dry climate.

What is *humification?*

Humification: process of transformation of plant tissues into humus.

What is *horizonation?*

Horizonation: degree or intensity of development of soil horizons.

CHAPTER 23—SAMPLE OBJECTIVE TEST QUESTIONS

A. MATCHING

1.	loam	a. _______	positive
2.	cutan	b. _______	moist
3.	cation	c. _______	dry
4.	acid	d. _______	texture
5.	udic	e. _______	pH = 5
6.	ustic	f. _______	organic
7.	humus	g. _______	clay skin

B. MULTIPLE CHOICE

1. The soil solum consists of the following soil horizons:

   ________a. A, E, B, and C.

   ________b. O, A, E and B.

   ________c. A, E and B.

   ________d. B and C.

2. Acid-generating cations in the soil solution include ions of

   ________a. aluminum, potassium, sodium.

   ________b. aluminum, hydrogen.

   ________c. hydrogen, potassium, calcium.

   ________d. hydrogen, magnesium, sodium.

3. Pedogenic processes consist of

   ________a. additions of material to the soil body.

   ________b. translocations of materials within the soil body.

   ________c. illuviation and eluviation.

   ________d. (all three of the above.)

C. COMPLETION

1. The smallest distinctive division of the soil of a given area is the ____________________; it is made up of ____________________.
2. Organic soil horizons are made up largely of ____________________, resulting from the pedogenic process of ____________________.
3. Clay minerals belong to a class of minerals called the ____________________ silicates.
4. In alkaline soils the large majority of cations held by soil colloids are ____________________ ____________________ cations.

CHAPTER 24

WORLD SOILS

EARLIER SOIL CLASSIFICATIONS

Where did modern soil science originate? Name two important soil scientists of the Russian school.

Who was responsible for the development of modern soil science in the United States?

THE COMPREHENSIVE SOIL CLASSIFICATION SYSTEM

By whom was the *Comprehensive Soil Classification System (CSCS)* developed?

State the important new concepts of the CSCS.

What is distinctive about the terminology of the CSCS?

THE SOIL TAXONOMY

What is the *Soil Taxonomy?*

PAGE

EARLIER SOIL CLASSIFICATIONS 458

Modern soil science founded by V. V. Dokuchaiev, 1880–1900; K. D. Glinka followed. Developed concepts of climate and vegetation control over horizons.

C. F. Marbut adapted Russian concepts, set up soil classification system within U. S. Dept. of Agriculture (USDA).

THE COMPREHENSIVE SOIL CLASSIFICATION SYSTEM 458

Comprehensive Soil Classification System (CSCS) was developed by USDA soil scientists, and soil scientists of university faculties and from foreign countries. System perfected in early 1970s.

Important new concepts of the CSCS: (1) defines classes strictly in terms of morphology and composition of the soils; (2) all definitions are quantitative; (3) definitions are in terms of observed or inferred features of the soil and are not subjective.

Large number of newly coined terms used in CSCS; terms contain syllables indicating nature of the soil or its environment.

THE SOIL TAXONOMY 458

Soil Taxonomy: the classification system of the CSCS, consisting of a hierarchy of six categories.

Name the six categories of the Soil Taxonomy. Approximately how many classes fall into each category?

Categories of the Soil Taxonomy:

Orders	10
Suborders	47
Great groups	185
Subgroups	1,000 (approx.)
Families	5,000 (approx.)
Series	10,000 (approx.)

DIAGNOSTIC HORIZONS FOR CLASSIFICATION

459

What are *diagnostic horizons?*

Diagnostic horizons: certain soil horizons, rigorously defined, that are used as diagnostic criteria in classifying soils within the Soil Taxonomy.

State the precise requirements of a *soil horizon,* as defined for the CSCS Soil Taxonomy.

Soil horizon, as defined for use in the CSCS Soil Taxonomy: (1) is a layer approximately parallel with the soil surface; (2) has a set of properties produced by soil-forming processes; (3) differs in properties from layers above and below it; (4) is differentiated from adjacent layers by characteristics that can be seen or measured in the field.

What are the two major groups of diagnostic soil horizons?

Two major groups of diagnostic soil horizons: (1) *epipedons;* (2) subsurface diagnostic horizons.

Define *epipedon.*

Epipedon: soil horizon that forms at the surface.

Epipedons

459

Describe the important epipedons listed below.

The following epipedons are of major importance in soil classification:

Mollic epipedon

Mollic epipedon: thick, dark-colored surface horizon rich in organic matter (humus) derived from roots or carried underground by animals; it is rich in base cations, PBS over 50%.

Umbric epipedon

Umbric epipedon: dark surface horizon, resembling mollic epipedon, but with PBS less than 50%.

Histic epipedon

Histic epipedon: thin surface horizon of peat, formed in wet places (bogs).

Ochric epipedon

Ochric epipedon: surface horizon light in color, with less than 1% organic matter.

Plaggen epipedon

Plaggen epipedon: Thick human-made surface layer produced by long-continued manuring, incorporating sod or other livestock bedding materials.

Subsurface Diagnostic Horizons

Describe the important subsurface diagnostic horizons listed below.

Argillic horizon

Agric horizon

Natric horizon

Calcic horizon

Petrocalcic horizon

Gypsic horizon

Plinthite

Salic horizon

Albic horizon

Spodic horizon

Cambic horizon

Oxic horizon

Other Horizons or Layers of Diagnostic Value

Describe the following diagnostic horizons:

Duripan

Subsurface Diagnostic Horizons 459

The following subsurface diagnostic horizons are important in the CSCS Soil Taxonomy:

Argillic horizon: an illuvial horizon (B horizon) in which layer-lattice clay minerals have accumulated by illuviation.

Agric horizon: illuvial horizon formed under cultivation and containing significant amounts of illuvial silt, clay, and humus, downwashed from plow layer.

Natric horizon: similar to argillic horizon, but with prismatic structure and high proportion of Na^+.

Calcic horizon: horizon of accumulation of calcium carbonate ($CaCO_3$) or magnesium carbonate ($MgCO_3$).

Petrocalcic horizon: hardened calcic horizon that does not break apart when soaked in water.

Gypsic horizon: horizon of accumulation of hydrous calcium sulfate (gypsum).

Plinthite: iron-rich concentrations in the form of dark red mottles in deeper horizons; capable of hardening into rocklike material with repeated wetting and drying.

Salic horizon: horizon enriched by soluble salts.

Albic horizon: pale, often sandy horizon (E) from which clay and free iron oxides have been removed.

Spodic horizon: horizon containing precipitated amorphous materials composed of organic matter and sesquioxides of Al, with or without Fe.

Cambic horizon: altered, fine-textured horizon that has lost sesquioxides or bases, including carbonates, through leaching; a variety of B horizon.

Oxic horizon: highly weathered horizon at least 30 cm (12 in.) thick, rich in clays and sesquioxides of low CEC; has few primary minerals capable of releasing base cations.

Other Horizons or Layers of Diagnostic Value 460

Other horizons of diagnostic value:

Duripan: dense, hard, subsurface horizon cemented by silica; does not soften during prolonged soaking.

Fragipan

Fragipan: dense, moderately brittle layer in which weak cementation is due to close packing and binding by clay.

Diagnostic Materials of Organic Soils

Describe the following diagnostic materials of organic soils:

Fibric soil materials

Hemic soil materials

Sapric soil materials

Diagnostic Materials of Organic Soils 460

Diagnostic materials of organic soils:

Fibric soil materials: organic matter consisting of fibers readily identifiable as to botanical origin.

Hemic soil materials: organic material of intermediated composition between fibric and sapric materials.

Sapric soil materials: highly decomposed organic matter with little content of identifiable fibers.

THE SOIL ORDERS

What are *soil orders?*

How are criteria for orders selected?

Group the ten soil orders into three broad classes and give a brief descriptive statement about each order.

Soils with poorly developed horizons:

Entisols

Inceptisols

Soils with a large proportion of organic matter:

Histosols

Soils with well-developed horizons:

Oxisols

Ultisols

THE SOIL ORDERS 460

Soil orders: those ten soil classes forming the highest category in the Soil Taxonomy of the CSCS.

Criteria of soil orders may include: (1) gross composition, whether organic or mineral or both; (2) degree of development of horizons; (3) presence or absence of certain diagnostic horizons; or (4) degree of weathering of soil minerals expressed as CEC or PBS.

Preliminary review of the ten soil orders, grouped into three broad classes based on degree of development or organic content:

Soils with poorly developed horizons or no horizons and capable of further mineral alteration:

Entisols: soils lacking horizons.

Inceptisols: soils having weakly developed horizons and containing weatherable minerals.

Soils with a large proportion of organic matter:

Histosols: soils with a thick upper layer of organic matter.

Soils with well-developed horizons or with fully weathered minerals, resulting from long-continued adjustment to prevailing soil-temperature and soil-water regimes:

Oxisols: very old, highly weathered soils of low latitudes, with an oxic horizon and low CEC.

Ultisols: soils or mesic and warmer soil-temperature regimes, with an argillic horizon and low base status (PBS $< 35\%$).

Vertisols

Vertisols: soils of subtropical and tropical zones with high clay content, developing deep, wide cracks when dry and showing evidence of movement between aggregates.

Alfisols

Alfisols: soils of humid and subhumid climates, with high base status (PBS > 35%) and an argillic horizon (B horizon).

Spodosols

Spodosols: soils with a spodic horizon (B horizon), an albic horizon (E horizon) with low CEC, and lacking carbonate minerals.

Mollisols

Mollisols: soils chiefly of midlatitudes, with a mollic epipedon and very high base status, associated with subhumid and semiarid soil-water regimes.

Aridisols

Aridisols: soils of dry climates, with or without argillic horizons, and with accumulations of carbonates or soluble salts.

Soil suborders

Soil suborders: second level of classification of soils in the Soil Taxonomy.

How are names of suborders formed?

Names of suborders combine two sets or syllables called formative elements.

What are formative elements?

Formative elements: syllables used to form the names of soil orders, suborders, and great soil groups in the Soil Taxonomy.

ENTISOLS

ENTISOLS 461

What are the *Entisols?*

Entisols have in common the combination of a mineral soil and the absence of distinct pedogenic horizons that would persist after normal plowing.

What are horizons lacking in the Entisols?

Horizons are lacking because parent material is unfavorable (e.g., quartz sand) or because of lack of time (e.g., recent alluvium, volcanic ash).

What global range have Entisols?

Entisols have wide global range: equatorial zone to arctic zone; included are glaciated areas, desert dune areas, floodplains, deltaic plains.

Comment on the agricultural potential of the Entisols.

Agricultural potential of Entisols: very poor in arctic zone and in deserts; very high on floodplains and deltaic plains of warm, moist climates.

Name the three important suborders of Entisols. Give a phrase or two to indicate the characteristics of each (see Appendix V).

Three important suborders of Entisols include:

Fluvents: Entisols formed from recent alluvium.

Orthents: Entisols on recent erosional surfaces, with shallow depth to bedrock regolith.

Psamments: Entisols with sandy texture, including dune sand and beach sand.

INCEPTISOLS

INCEPTISOLS 462

What are the five properties of the *Inceptisols?*

Inceptisols are uniquely defined by these properties: (1) soil water is readily available in part of each year; (2) horizons formed by alteration or concentration are present, but without accumulation of translocated materials other than carbonates or silica; (3) textures are finer than loamy sand; (4) soil has some weatherable minerals; (5) clay fraction has moderate to high CEC.

Over what range of latitudes are the Inceptisols found?

Inceptisols are found in wide latitude range, e.g., tundra, high mountains at all latitudes.

On what kinds of surfaces are Inceptisols found?

Inceptisols occur on relatively young landform surfaces, e.g., glaciated areas, alluvial plains (low floodplain terraces). Soil is young, but has had time to develop horizons.

Of the six suborders of Inceptisols, name the one most important.

The one most important suborder of Inceptisols is:

Aquepts: Inceptisols of wet places.

HISTOSOLS

HISTOSOLS 470

What are the *Histosols?*

Histosols: soils having a very high content of organic matter in the upper 80 cm (32 in.); most are peats or mucks.

What are the properties of Histosols of cool and cold climates?

Histosols of cool or cold climates are acid and low in plant nutrients; widespread in midlatitude, subarctic, and arctic zones.

In what way are Histosols potentially good agricultural soils?

Histosols of midlatitudes, when drained and treated with lime and fertilizers, can become highly productive for garden crops. Peat is used as garden mulch or as low-grade fuel.

OXISOLS

OXISOLS 471

What are the three unique properties of *Oxisols?*

Oxisols are unique through three properties: (1) extreme weathering of most minerals to sesquioxides of Al and Fe and to kaolinite; (2) very low CEC of the clay fraction; (3) a loamy or clayey texture.

What diagnostic horizon is usually present in Oxisols?

Oxic horizon is usually present within 2 m (6 ft.) of surface. Plinthite may occur in deeper parts of profile.

Where and under what conditions have the Oxisols developed?

Oxisols are developed in equatorial, tropical, and subtropical zones on land surfaces stable over long periods of time (Pleistocene or much older). Soil must be moist when developed.

Are horizons well developed in Oxisols? What color are Oxisols?

Distinct horizons are lacking. Colors are red, yellow, yellowish-brown.

How well are Oxisols supplied with nutrients?

Oxisols are low in nutrients. Addition of fertilizers may be ineffective because phosphorus is held in unavailable forms.

Describe the physical structure of the *Oxisols.*

Oxisols are friable; easily penetrated by plant roots.

With what climates are the Oxisols most closely associated?

Oxisols are closely associated with wet equatorial climate (1) and wet-dry tropical climate (2) in equatorial and tropical zones of South Africa and South America.

What agricultural used are made of Oxisols? Explain.

Oxisols are used largely for shifting (slash-and-burn) agriculture. Small nutrient supply is in shallow surface layer; is quickly exhausted.

ULTISOLS

ULTISOLS 472

What are the unique properties of the *Ultisols?*

Ultisols are unique through the combination of three properties: (1) an argillic horizon; (2) supply of exchangeable bases (CEC) is low; (3) mean annual soil temperature is greater than 8°C.

Where is PBS highest in the Ultisols? Where lowest?

In Ultisols, PBS is highest close to surface, diminishes rapidly with depth.

Under what conditions have the Ultisols formed?

Ultisols form under forest in moist climates (udic regime) and subhumid climates (ustic regime). Surplus water causes leaching of bases.

What is distinctive about the B horizon of the Ultisols?

B horizon is red or yellowish-brown, because of concentration of sesquioxides of Fe.

What is the global distribution of the Ultisols in terms of climate?

Ultisols important in moist subtropical climate (6), in wet-dry tropical climate (3), in monsoon and trade-wind littoral climate (2).

What is the history of the Ultisols?

Ultisols have formed on land surfaces long subject to erosion and weathering; often with thick saprolite (residual regolith).

What agricultural use is made of the Ultisols?

Ultisols of low latitudes used in shifting (slash-and-burn) agriculture. Will respond to treatment with lime and fertilizers, as in S. E. United States, S. China.

VERTISOLS

VERTISOLS 472

What are the unique properties of the *Vertisols?*

Vertisols are uniquely defined through these properties: (1) a high content of clay (montmorillonite) that shrinks and swells greatly with changes in soil-water storage; (2) deep, wide cracks at some season; (3) evidences of soil movement.

What features give evidence of soil movement in the Vertisols?

Evidences of soil movement are *slickensides* and *gilgae;* also heaving of soil-supported objects.

Define *slickensides.*

Slickensides: soil surfaces with grooves or striations made by movement of one mass of soil against another while the soil is moist.

Define *gilgae.*

Gilgae: small relief features that may be knobs and basins, or narrow ridges with intervening troughs.

With what climate and vegetation are Vertisols associated?

Vertisols are associated with the semiarid subtype of the dry tropical climate (4s) and the wet-dry tropical climate (3). Natural vegetation is grassland or savanna.

Explain how alternation of wet and dry seasons causes changes in the Vertisols.

In wet (rainy) season, soil fragments fall into deep cracks; cracks then close, "swallowing" fragments. In dry season, wide cracks open up as soil shrinks.

What are some other names for Vertisols throughout the world?

Vertisols also called *black cotton soils, tropical black clays,* and *regur.*

What is the base status of the Vertisols? Are they acid or alkaline?

Base status of Vertisols is very high; soil is neutral in pH.

What can be said of the water-holding properties of the Vertisols?

Vertisols have high capacity to hold soil water; but much is held by clay minerals and not available to plants. Soil when moist is tough, sticky, difficult to till. Cannot be cultivated by primitive methods.

What are the prospects for development of agriculture on Vertisols?

Prospects for agriculture may be excellent with use of modern machinery and soil treatment, requiring large input of cultural energy.

ALFISOLS

ALFISOLS 473

What are the unique properties of the *Alfisols?*

Alfisols are uniquely defined by the following properties: (1) a gray, brownish, or reddish horizon (ochric epipedon) not darkened by humus; (2) a horizon of clay accumulation (argillic horizon); (3) medium to high PBS; and (4) soil water available to plants over a large part of the year.

Describe the argillic horizon.

Argillic horizon of the Alfisols is a B horizon, enriched by accumulated silicate clay minerals and moderately saturated with exchangeable bases (Ca, Mg).

What characterizes the E horizon?

The E horizon shows some loss of bases, silicate clay minerals and sesquioxides.

With what climates are the Alfisols associated?

Alfisols are dominantly in areas of the moist continental climate (10), Mediterranean climate (7) and marine west-coast climate (8) in North America, Europe, Asia, and Australia. Alfisols also occur with the wet-dry tropical climate (3) and the semiarid subtype of dry tropical and subtropical climates (4s, 5s).

Are Alfisols agriculturally productive? Explain.

Alfisols are fairly productive agriculturally under simple management; usually have adequate soil water and nutrient bases, but benefit from use of fertilizers.

Of the five suborders of Alfisols, four deserve recognition. Name and describe them briefly.

Four of the five suborders of Alfisols are:

Boralfs: Alfisols of boreal forests or high mountains, associated with mean annual soil temperature less than 8°C.

Udalfs: Alfisols of the ustic soil-water regime, brownish to reddish in color.

Ustalfs: Alfisols of the ustic soil-water regime, brownish to reddish in color.

Xeralfs: Alfisols of the xeric coil-water regime in the Mediterranean climate (7).

SPODOSOLS

SPODOSOLS 474

What is the unique property of the *Spodosols?*

Spodosols are uniquely defined by presence of the spodic horizon of accumulation of dark, amorphous materials.

Describe the spodic horizon.

Spodic horizon, a B horizon, has amorphous materials consisting of organic material and compounds of Al and commonly Fe, brought downward by illuviation from the overlying E horizon.

Are Spodosols acid or alkaline? High or low in base cations?

Spodosols are strongly acid, low in nutrient base cations (Ca, Mg), low in humus.

What texture is associated with the Spodosols?

Spodosols are closely associated with sand texture, have low water-holding capacity. Parent material lacks clay minerals.,

With what climate and vegetation are the Spodosols associated?

Spodosols are formed largely under forest cover in the boreal forest climate (11) and colder parts of the moist continental climate (10). Forest is mostly needleleaf (boreal type).

How are areas of Spodosols related to areas of continental glaciation?

Spodosol areas correspond closely with areas of most recent Pleistocene glaciation. (Spodosols also occur in lower latitudes.)

Evaluate the Spodosols in terms of agricultural productivity.

Spodosols are naturally poor agricultural soils; need heavy lime and fertilizer applications. Growing season is very short in northerly parts.

MOLLISOLS

What are the unique properties of the *Mollisols?*

Mollisols are uniquely defined through combination of these properties: (1) a mollic epipedon is present; (2) calcium is dominant among the base cations in the A and B horizons; (3) a dominance of crystalline clay minerals of moderate or high CEC; (4) less than 30% clay in some horizon if the soil has deep, wide cracks in some season.

Describe the mollic epipedon of the Mollisols.

Mollic epipedon is a very dark brown to black surface horizon that is more than one-third of the combined thickness of the A and B horizons or is more than 25 cm thick. It has a soft consistence when dry.

Under what conditions of climate and vegetation do the Mollisols form?

Mollisols form under grass cover in climates with a pronounced seasonal soil-water deficiency, i.e., semiarid subtype of the dry midlatitude climate (9s) and the subhumid subtypes of the moist continental climate (10h and 10sh) and moist subtropical climate (6sh).

Where are the largest regions of occurrence of the Mollisols?

Largest regions of occurrence of Mollisols: Great Plains of N., America, Eurasian steppes; Pampas of S. America.

How fertile are the Mollisols? Explain.

Mollisols have very high natural fertility for grains and grasses. Well-developed granular texture favors tillage. Base status is very high.

What agricultural products are most important in areas of Mollisols?

Mollisols are source regions of most of grain (esp. wheat) in commercial trade. Used by primitive nomadic societies for grazing.

Of the seven suborders of Mollisols, four deserve comment. Name and describe them briefly.

Of the seven suborders of Mollisols the four that deserve comment include:

Borolls: Mollisols of cold-winter semiarid plains (steppes) or high mountains. Mean annual soil temperature is below 8°C.

Udolls: Mollisols of the udic soil-water regime; they have no horizon of $CaCO_3$ accumulation. Associated with tall-grass prairie.

Ustolls: Mollisols of the ustic soil-water regime. Most have $CaCO_3$ accumulation in C horizon (*caliche*).

Xerolls: Mollisols of the xeric regime under the Mediterranean climate (7).

ARIDISOLS

What are the unique properties of the *Aridisols?*

ARIDISOLS 476

Aridisols are uniquely defined by the following combination of properties: (1) lack of water available for plants for very extended periods; (2) one or more pedogenic horizons; (3) surface horizon not darkened by humus; (4) absence of deep, wide cracks.

Under what conditions do Aridisols form?

Aridisols form in climates with scanty rainfall: semidesert and desert subtypes of all dry climates (4sd, 4d, 5sd, 5d, 9sd, 9d) and Mediterranean climate (7s, 7sd).

What is the prospect for agricultural production on the Aridisols?

Aridisols can be highly productive under irrigation because of high content of base cations.

Name and describe briefly the suborders of Aridisols (see Appendix V).

Suborders of Aridisols;

Argids: Aridisols with an argillic horizon, mostly formed on older land surfaces and probably influenced by earlier moist (pluvial) climatic periods.

Orthids: Aridisols without an argillic horizon but with one or more pedogenic horizons. May have horizon of salt accumulation (salic horizon).

SOILS AND ALTITUDE

How does the soil profile change with an increase in altitude? Give an example from western U. S.

SOILS AND ALTITUDE 477

With an increasing altitude, soil profile changes to reflect cooler, moister climate. Example from western U. S.: succession from Aridisols on arid basin floors, through Mollisols (Ustolls, Udolls) on rising mountain slopes, to Spodosols on summits.

THE CANADIAN SYSTEM OF SOIL CLASSIFICATION 478

Why was if desirable to develop a Canadian system of soil classification?

Special needs of a workable Canadian soil classification system: to accommodate vast areas of boreal forest and tundra with parent materials recently deposited by continental glaciers; to deal only with soils within Canadian boundaries.

What is the overall philosophy behind the Canadian system?

Overall philosophy: a reasonable and usable system with taxa based on properties of the soils themselves.

How are classes defined under the Canadian system?

Classes are defined on kinds, degrees of development, and sequence of soil horizons.

SOIL HORIZONS AND OTHER LAYERS

What are the major mineral horizons?

What are the major organic horizons?

How is the *A horizon* defined?

How is the *B horizon defined?*

How is the *C horizon* defined?

How are subdivisions of the these horizons indicated?

SOIL HORIZONS AND OTHER LAYERS 478

Major mineral horizons: A, B, and C.

Major organic horizons: L, F, H, and O.

A horizon: mineral horizon, at surface, in zone of leaching or eluviation.

B horizon: mineral horizon enriched in organic matter, sesquioxides, or clay.

C horizon: mineral horizon unaffected by pedogenic processes.

Subdivisions are indicated by lowercase suffixes.

SOIL ORDERS OF THE CANADIAN SYSTEM

How many soil orders make up the Canadian system? Name them.

SOIL ORDERS OF THE CANADIAN SYSTEM 479

The nine soil orders that make up the Canadian system include: *Brunisolic, Chernozemic, Cryosolic, Gleysolic, Luvisolic, Organic, Podzolic, Regosolic, Solonetzic.*

BRUNISOLIC ORDER

What is the central concept of the *Brunisolic order?*

Under what conditions do Brunisolic soils occur?

Compare the B horizon of the Brunisolic soil with that of the Chernozemic soils and the Podzolic soils.

BRUNISOLIC ORDER 479

Brunisolic order: soils under forest having brownish-colored Bm horizon (horizon slightly altered by hydrolysis, oxidation, or solution).

Brunisolic soils occur in areas of boreal forest, mixed forest, shrubs, grass, heath, or tundra.

Brunisolic B horizon is weak; Chernozemic soils lack a B horizon. Podzolic soils have strongly developed B horizon of accumulation.

CHERNOZEMIC ORDER

What is the concept of the *Chernozemic order?* Describe the A horizon.

Describe the C horizon.

CHERNOZEMIC ORDER 480

Chernozemic order: imperfectly drained soils with thick, dark surface horizon (organic), called Ah horizon.

C horizon has lime accumulation (Cca).

CRYOSOLIC ORDER

Under what environmental conditions is the *Crysolic order* found?

Are organic horizons present?

CRYOSOLIC ORDER 481

Cryosolic order: predominates in permafrost region north of the tree line; here cryoturbation is strong.

Organic horizons (L, H, and O) are present.

GLEYSOLIC ORDER

Describe soils of the *Gleysolic order.*

Describe the Ah horizon.

GLEYSOLIC ORDER 481

Gleysolic order: soils showing features indicative of periodic or prolonged water saturation and reducing conditions.

The Ah horizon is thick, originating in hygrophytic vegetation.

LUVISOLIC ORDER

What features characterize the *Luvisolic order?*

Under what condition do these soils develop?

LUVISOLIC ORDER 482

Luvisolic order: soils with light-colored eluvial horizon (Ae) and illuvial B horizons (Bt) with accumulated silicate clay.

Luvisolic soils develop in base-saturated parent materials in imperfectly drained sites, in mild to very cold climates.

ORGANIC ORDER

Describe soils of the *Organic order.*

Under what conditions do they occur?

ORGANIC ORDER 482

Organic order: soils largely of organic materials; they are known as peat, much, or bog soils.

Organic soils occur in poorly drained depressions in humid to perhumid climates.

PODZOLIC ORDER

What is distinctive about soils of the *podzolic order?*

With what texture of parent material are Podzolic soils associated?

Is an organic horizon present?

Describe the B horizon.

PODZOLIC ORDER 482

Podzolic order: soils having B horizons of dominant accumulation of amorphous material composed mainly of humified organic matter, along with aluminum and iron.

Texture of acid parent materials is coarse to medium, occurring under forest or heath.

Podzolic soils have organic surface horizons, commonly L, F, and H.

B horizon is reddish brown to black (Bh) with abrupt boundary.

REGOSOLIC ORDER

What characterizes soils of the *Regosolic order?*

REGOSOLIC ORDER 482

Regosolic order: soils with weakly developed horizons because of youthfulness of the parent material or instability of slopes.

SOLONETZIC ORDER

What is the nature of soils of the *Solonetzic order?*

SOLONETZIC ORDER 482

Solonetzic order: soils with B horizons that are very hard when dry and swell to a sticky mass when wet.

Describe the B horizon.	B horizon has prismatic or columnar structure that breaks into hard, blocky peds.
Under what conditions do Solonetzic soils occur?	Solonetzic soils occur on saline parent materials.
What is *solodization?*	Solonetzic soils undergo a development program leading to *solodization,* in which the B horizon is broken down and destroyed.

CHAPTER 24—SAMPLE OBJECTIVE TEST QUESTIONS

A. MATCHING

1.	epipedon	a. _______	gilgae
2.	plinthite	b. _______	plowed
3.	plaggen	c. _______	Ultisols
4.	fibric	d. _______	at surface
5.	Vertisols	e. _______	lacking horizons
6.	argillic horizon	f. _______	red mottles
7.	Entisols	g. _______	peat

B. MULTIPLE CHOICE

1. Which of the following is not a diagnostic horizon:

   ________a. umbric epipedon.

   ________b. agric horizon.

   ________c. hemic horizon.

   ________d. plinthite.

2. Which of the following is <u>not</u> a diagnostic property of the Oxisols?

   ________a. low CEC of the clay fraction.

   ________b. extreme weathering of most minerals.

   ________c. a mollic epipedon.

   ________d. an oxic horizon.

3. Which of the following is <u>not</u> a diagnostic property of the Ultisols?

   ________a. a spodic horizon.

   ________b. an argillic horizon.

   ________c. CEC is low in lower horizons.

   ________d. mean annual soil temperature is below 8°C.

C. COMPLETION

1. A soil horizon that forms at the surface is an ____, an example of an epipedon caused by cultivation is a ____ epipedon.

2. A subsurface horizon rich in sodium is a ____ horizon; one rich in soluble salts is a ____ horizon.

3. Texture of the spodic horizon is dominated by ____ .
4. Soil horizons, rigorously defined, that are used as diagnostic criteria in classifying soils within the Soil Taxonomy are known as the ____ .

CHAPTER 25

ENERGY FLOWS AND MATERIAL CYCLES IN THE BIOSPHERE

PAGE

Define *biosphere.*

Biosphere: all living organisms of the earth and the environments with which they interact.

Define *ecology.*

Ecology: science of interactions between life forms and their environment. The science of ecosystems.

Define *ecosystem.*

Ecosystem: the total assemblage of components entering into the interactions of a group of organisms.

What concept relates ecosystems with energy and matter?

Concept: ecosystems depend upon inputs and outputs of energy and matter to maintain their biological structures.

How does the geographer view ecosystems?

Concept: Geographer views ecosystems as natural resource systems.

THE ECOSYSTEM AND THE FOOD CHAIN

THE ECOSYSTEM AND THE FOOD CHAIN 482

Define *food chain.*

Food chain: organization of an ecosystem into steps or levels through which energy flows as organisms at each level consume energy stored in the bodies of organisms of the next lower level.

Define *primary producers.* To what group of organisms do most primary producers belong?

Primary producers: organisms which can convert carbon dioxide and water with light energy to photosynthesis into carbohydrates. Primary producers are at bottom level of food chain. Most primary producers are green plants.

Define *consumers.*

Consumers: animals in the food chain that live on organic matter formed by primary producers or by other consumers.

Define *primary consumers.*

Primary consumers: animals that live by feeding on producers.

Define *secondary consumers.*

Secondary consumers: animals that feed upon primary consumers.

Define *decomposers.* To what groups of organisms do most decomposers belong? On what do they feed?

Decomposers: organisms that feed on dead organisms from all levels of food chain. Most consumers are microorganisms and bacteria which feed on decaying organic matter.

What concept relates solar energy to stored energy in the food chain?

Concept: solar energy is converted by primary producers into stored energy in form of carbohydrate compounds used in turn as energy source for successive levels of consumers in food chain.

PHOTOSYNTHESIS AND RESPIRATION

PHOTOSYNTHESIS AND RESPIRATION 485

Define *photosynthesis.*

Photosynthesis: production of carbohydrate by the union of water with carbon dioxide while absorbing light energy.

Define *carbohydrate.*

Carbohydrate: class of organic compounds consisting of the elements carbon (C), hydrogen (H), and oxygen (O).

Describe the chemical nature of the carbohydrate molecule.

Carbohydrate molecule is composed of chains of C atoms, H atoms, and hydroxyl atom-pairs (OH) symbolized as —CHOH—.

What compounds enter into photosynthesis? Show the chemical reaction. What gas is a byproduct?

Photosynthesis involves water (H_2O) and carbon dioxide (CO_2) in chemical reaction (simplified):

$$H_2O + CO_2 + \text{light energy} \longrightarrow \text{—CHOH—} + O_2$$

Byproduct is oxygen gas (O_2).

Define *respiration.* What form of energy sustains the organism? What form of energy is lost? Where does it go?

Respiration: a metabolic process in which organic compounds are oxidized within living cells to yield biochemical energy and waste heat. The biochemical energy sustains the organism, while the heat energy is lost to the outside and thus leaves the ecosystem.

Give a simplified chemical reaction for respiration.

Respiration:

$$\text{—CHOH—} + O_2 \longrightarrow CO_2 + H_2O + \text{chemical energy}$$

What happens to the chemical energy released by respiration?

Chemical energy released in respiration may be stored in energy-carrying molecules for later use in life processes. Ultimately energy is released to environment.

Organize the photosynthesis-respiration reactions into a cycle, or loop. Include producer and consumer in the loop. Trace carbon dioxide, water and oxygen through the loop. Add the flow of solar energy and longwave radiation to the system.

Photosynthesis-respiration reactions form a cycle, or loop. Carbon dioxide and water are taken in from atmosphere and soil by producer, then released to atmosphere by decomposer. Oxygen is liberated to the atmosphere in photosynthesis, then absorbed from the atmosphere in respiration. Light energy of solar origin absorbed in the system is ultimately liberated as longwave radiation and eventually lost to space.

What concept contrasts the flows of energy and matter in the photosynthesis-respiration system? Is recycling involved?

Concept: energy enters and leaves the global photosynthesis-respiration system, whereas material components are simply recycled.

NET PHOTOSYNTHESIS

NET PHOTOSYNTHESIS 486

Define *gross photosynthesis.*

Gross photosynthesis: total amount of carbohydrate produced by photosynthesis by a given organism or group of organisms in a given unit of time.

Define *net photosynthesis.*

Net photosynthesis: carbohydrate remaining after respiration has broken down sufficient carbohydrate to power the metabolism of the organism.

Give the equation for net photosynthesis, using words.

In equation form:

Net photosynthesis = Gross photosynthesis − Respiration

How can net photosynthesis be used as an indicator?

New photosynthesis can be measured and used as as indicator of intensity of primary productivity.

How does net photosynthesis change with an increase in light intensity? Why does a decline set in?

Net photosynthesis and available light energy: rate of net photosynthesis first rises rapidly with increasing light intensity, then levels off, reaching a plateau, then declines. Decline may result from increased respiration rate at higher temperature.

In what way does net photosynthesis depend upon duration of daylight? Upon what factors does duration of daylight depend? How does daylight depend upon latitude?

Net photosynthesis and duration of daylight: daily amount of net photosynthesis depends upon number of hours of daylight, which depends upon latitude and season. Near equator, daylight lasts about 12 hours throughout the year. At high latitudes long daylight period occurs in summer only. Daylight duration can be read from graph.

How does net photosynthesis change with an increase in air temperature? At what temperature is the peak reached? What causes a decline in rate of net photosynthesis?

Net photosynthesis and air temperature: net photosynthesis increases rapidly with rising temperature peaks at about 18°C (65°F), then falls off rapidly because of increase in respiration rate.

ENERGY FLOW ALONG THE FOOD CHAIN

ENERGY FLOW ALONG THE FOOD CHAIN 487

What concept relates the loss of energy to upward steps in the food chain? What percentage of energy is lost in each step? How does energy loss limit the number of steps?

Concept: energy is lost with each step upward in the food chain. Energy passed upward may range from 10% to 50%. Energy loss limits the numbers of levels in the food chain.

In terms of world food resources, what advantage lies in consumption of plant foods rather than meat? Explain.

Significance in world food resources: Reduction in human consumption of meat in advanced nations in favor of increased consumption of plant foods (grains, legumes) would increase the food resource for a given input of fossil fuel energy.

THE ENERGY FLOW SYSTEM OF A GREEN PLANT

THE ENERGY FLOW SYSTEM OF A GREEN PLANT 487

Describe a model energy flow system of a green plant.

Input of energy flow system of a green plant is solar shortwave energy, which is converted to chemical energy by photosynthesis, initially as *labile energy,* then by respiration to *biomass energy.* Energy leaves the system at various points as sensible heat, latent heat, and longwave radiation.

What is *labile energy?*

Labile energy: chemical energy that is continually undergoing change by respiration and conversion into chemical energy stored in more complex molecules.

What is *biomass energy?*

Biomass energy: chemical energy stored in the tissues of plants in the form of various kinds of organic molecules.

How can the food chain be shown by a similar schematic diagram?

Food chain can be shown as succession of energy flow systems, in abbreviated symbol, passing from primary producer to first level of consumer, and so forth to higher levels in the chain.

NET PRIMARY PRODUCTION

NET PRIMARY PRODUCTION 488

Define *biomass.* What units of measure are used?

Biomass: the dry weight of living organic matter in an ecosystem within a designated surface area, i.e., grams of organic matter per square meter.

In what organic systems do we find the largest biomass? Why?

Forests have largest biomass because so much carbohydrate is stored in woody tissues.

Define *net primary production.* What units of measure are used?

Net primary production: rate at which carbohydrate is accumulated in the tissues of plants within a given ecosystem. Stated in units of grams of dry organic matter per year per square meter of surface.

In what environments are values highest? intermediate in value? lowest?

Values are highest in rainforest, freshwater swamps and marshes, tidal estuaries. Intermediate values in grasslands, agricultural lands, lakes and streams, and continental shelf. Low values in desert and open ocean.

What concept relates net primary production of the oceans to depth of water and position with respect to continents? What effect has upwelling upon net primary production? Explain.

Concept: net primary productivity of oceans lies largely in shallow waters of continental shelves and in narrow zones of upwelling. Upwelling brings up nutrients to intensify growth of phytoplankton as base of food chain. Example: Peru Current and anchoveta.

NET PRODUCTION AND CLIMATE

NET PRODUCTION AND CLIMATE 490

Upon what factors does the rate of net primary production depend?

Net primary production rate depends upon seasonal cycles and values of light intensity, temperature, and soil-water balance.

Describe the correlation of net primary production and climate type. Name the climates for which production has a rating in the following scale: highest, high, moderate, low, very low.

Close correlation of net primary production and climates. Highest in low-latitude wet, and wet-dry climates. High in wet-dry tropical, moist subtropical, and marine west-coast climates. Moderate in Mediterranean and moist continental climates. Low in steppes and boreal forest climates. Very low in deserts and tundra.

BIOMASS ENERGY 493

How can biomass energy be used as an energy resource?

Biomass energy makes use of the chemical energy stored in plant tissues through photosynthesis. Direct burning of tissues (wood) and combustion of intermediate fuels derived from plant matter (gas, charcoal, alcohol) are forms of biomass energy.

Define *biogas?* How is it used?

Biogas: mixture of methane and carbon dioxide produced by bacterial digestion of human and animal waste. It may be burned for cooling or heating.

How is alcohol produced from biomass? How is it used?

Alcohol is produced for biomass by bacterial fermentation of yeast. Alcohol can be used alone or combined with gasoline for motor fuel use.

What impact does burning of biomass to release CO_2 have on the environment?

There is no gain in CO_2 because the biomass would decay to produce CO_2 naturally in a short period of time. If the biomass energy replaces fossil fuel burning, then CO_2 production by this means is reduced.

MATERIAL CYCLES IN ECOSYSTEMS

MATERIAL CYCLES IN ECOSYSTEMS 493

Compare a material cycle with an energy cycle. Which cycle is open and which is closed?

Energy cycles are open cycles (open systems); energy enters and leaves the global energy system *Material cycles* are closed cycles (closed systems) in which matter remains earthbound and cannot be lost to space.

Define *material cycle.*

Material cycle: total system of pathways by which a particular type of matter (element, compound, ion) moves through the earth's ecosystem or biosphere (also called *biogeochemical cycle* and *nutrient cycle*).

What is a *sedimentary cycle.*

Sedimentary cycle: type of material cycle in which the compound or element is released from rock by weathering, follows the movement of running water either in solution or as sediment to reach the sea and is eventually converted into rock.

What is a *gaseous cycle?*

Gaseous cycle: type of material cycle in which an element or compound is converted to gaseous form, diffuses through the atmosphere, and passes rapidly over land or sea, where it is reused in the biosphere.

Give examples of substances that move in gaseous cycle.

Primary constituents of living matter (C, H, O, N) move in gaseous cycles.

What is a *pool* of materials?

Pool of materials: area or location of concentration of a given material in the material cycle.

What two kinds of pools are there?

In *active pools,* materials are in forms and places readily accessible to life processes. In *storage pools,* materials are more or less inaccessible (for example, in carbonate sediments).

NUTRIENT ELEMENTS IN THE BIOSPHERE

NUTRIENT ELEMENTS IN THE BIOSPHERE 494

Define *macronutrients.* Name the top three macronutrients, giving percentages of abundance for each. Which macronutrient rates third, and in what percentage? List the remaining macronutrients.

Macronutrients: nine elements required in greatest abundance for organic growth, including primary production by green plants. Top three are hydrogen, almost 50%; carbon, almost 25%; oxygen, almost 25%. Remaining 0.5% includes nitrogen, calcium, potassium, magnesium, sulfur, and phosphorus.

THE CARBON CYCLE

THE CARBON CYCLE 494

What is the *carbon cycle?*

Carbon cycle: a material cycle in which carbon moves through the biosphere; it includes both gaseous cycles and sedimentary cycles.

How does carbon move in the gaseous portions of the cycle?

Carbon moves as CO_2 gas in atmosphere and hydrosphere.

How does carbon move in the sedimentary portions of the cycle?

Carbon moves as $CaCO_3$ in sedimentary portions of cycle; also moves as hydrocarbon compounds (coal, petroleum, natural gas).

What biologic processes make use of carbon and release carbon?

In biologic processes, carbon is used by plants in photosynthesis and released to the atmosphere in respiration.

What impact have humans on the carbon cycle?

Humans enter the carbon cycle through combustion of fuels, releasing CO_2 into the atmosphere.

THE OXYGEN CYCLE

What is the *oxygen cycle?*

How does oxygen move in the gaseous portions of the cycle?

How does oxygen move in the biological portions of the cycle?

Name three ways that human activity reduces the amount of oxygen in the air.

THE OXYGEN CYCLE 495

Oxygen cycle: a material cycle in which oxygen moves through the biosphere in both gaseous and sedimentary cycles.

Oxygen moves as molecular oxygen gas (O_2) in the gaseous portion of the cycle within the atmosphere and hydrosphere. (Oxygen also moves as CO_2 gas, as explained under carbon cycle.)

Oxygen is released to atmosphere by plant photosynthesis; is taken up from atmosphere in respiration.

Three ways human activity reduces the amount of oxygen in the air include: (1) burning fossil fuels; (2) clearing and draining land which speeds oxidation of soils and soil organic matter; (3) reducing photosynthesis by clearing forests for agriculture and paving previously productive surfaces.

THE NITROGEN CYCLE

What is the *nitrogen cycle?*

What is *nitrogen fixation?* How is it carried out?

What are the symbiotic nitrogen fixers?

What is *denitrification?*

How do humans influence the nitrogen cycle?

THE NITROGEN CYCLE 496

Nitrogen cycle: material cycle in which nitrogen moves through the biosphere by processes of *nitrogen fixation* and *denitrification.*

Nitrogen fixation: chemical process of conversion of gaseous molecular nitrogen (N_2) or atmosphere into compounds or ions that can be directly used by plants. Certain microorganisms (soil bacteria, blue-green algae) can fix nitrogen directly.

Symbiotic nitrogen fixers, bacteria of the genus *Rhizobium,* live in physical contact with certain plants (example, legumes). Nitrogen-fixing plants are important in nitrogen cycle.

Denitrification: biochemical process in which nitrogen in forms usable to plants is converted into molecular nitrogen (N_2) in gaseous form and released to atmosphere.

Humans influence nitrogen cycle through synthetic fixation for fertilizers; rate now equals natural biological fixation rate. Nitrogen, as nitrate ions, stimulates plant growth in rivers and lakes.

SEDIMENTARY CYCLES

Sketch a flow diagram of nutrient cycling through producers, consumers, and decomposers. Indicate storage in soil and soil colloids; major storage pools in seawater and bedrock.

SEDIMENTARY CYCLES 497

Cycling of nutrients represented by flow diagram (Figure 25.17 pg. 497). Flow paths pass through producers, consumers, decomposers. Storage in soil on soil colloids. Major storage pools in seawater and bedrock.

AGRICULTURAL ECOSYSTEMS

AGRICULTURAL ECOSYSTEMS 498

Give three major differences between natural and agricultural ecosystems.

Differences between natural and agricultural ecosystems: (1) reliance of agricultural ecosystems on inputs of fossil fuel energy. Fuel, fertilizers, pesticides, transportation; (2) simplicity of agricultural ecosystems; usually one genetic strain of one plant species. Sensitive to insect predators, disease. Weeds must be controlled in agricultural ecosystems; (3) fertilizers used in agricultural ecosystems are removed with harvest, not recycled.

PRODUCTIVITY AND EFFICIENCY OF AGRICULTURAL SYSTEMS

PRODUCTIVITY AND EFFICIENCY OF AGRICULTURAL SYSTEMS 498

By what factor can net primary productivity be increased within agricultural ecosystems by use of fossil fuel inputs.?

Boost in net primary productivity in agricultural ecosystems: fossil fuel energy input can raise productivity by factor of 5 to 10 over primitive agriculture.

In what two major categories is energy consumed in agriculture?

Two major categories of energy consumption in agriculture: (1) petroleum and electricity expended on farms; (2) energy represented by materials expended and equipment used on farms.

What percentage of the total energy expenditure of annual agricultural production on farms is in the form of fertilizers?

Fertilizers account for about 30% of the total energy expenditure.

Define *cultural energy.*

Cultural energy: energy in forms exclusive of solar energy of photosynthesis that is expended on production of raw food or feed crops.

Name some food and feed crops that are highly efficient in terms of cultural energy expended. Name some crops of low efficiency.

Crops highly efficient in terms of cultural energy demands: sorghum, corn, wheat, sugarcane, soybean, oats. Crops of low efficiency: fruits, garden vegetables.

Which agricultural products give the highest protein content in terms of cultural energy expended?

High protein yield per unit of cultural energy: alfalfa, soybean, sorghum. Low yields: peanuts, rice, potato, garden vegetables. Very low yield: chicken, beef, pork.

THE GREEN REVOLUTION—SUCCESS OR FAILURE? 500

What is the *green revolution*? What does it include?

Green revolution: program of strategies and technologies to increase production of crops in Third World countries. It includes development of new genetic strains of crops, changes in agricultural practices, and increased inputs of fertilizer and pesticides.

What countries are affected by the green revolution?

The countries affected by the green revolution are in low-latitude, climates with strong wet-dry seasons.

Who was Norman Borlaug? What was his contribution to the green revolution? Where was it applied?

Norman Borlaug was a agronomist who received the Nobel Peace Prize for developing a high-yielding dwarf wheat variety. The new variety was used very successfully in Mexico, Pakistan, and India.

What other crop was improved? Where?

New, high-yielding strains of rice were developed in the Philippines and applied there.

What problems developed in the early years of the green revolution?

The new strains required dependable supplies of water, fertilizer, pesticides, and financial credit, which were often not available to poor farmers.

What important factor counteracted green revolution crop increases?

Crop increases were offset by increases in population that consumed the added production.

What was the result in Asia? In Africa?

In Asia, agricultural productivity increased to the point of self-sufficiency for most countries. In Africa, important advances, such as the introduction of drought-resistant sorghum, were made.

What are three of the drawbacks of the use of new generic strains of crops?

The three drawbacks of the use of new generic strains of crops include: (1) a single strain is vulnerable to disease, so wide application may be risky; (2) field sizes need to be increased and so fewer types of crops are grown, making small farmers reluctant to rely on a single strain; (3) crops of a single strain may be destroyed by insect predation, in spite of pesticides.

How can future increases be achieved?

Future increases can be achieved only by making new strains, cultivation, and fertilization methods more compatible with existing agricultural systems and local culture patterns.

Can land area under cultivation be increased?

Although some large areas of vertisols could be placed under cultivation with expensive mechanical cultivation, major expansion of the world's cultivated lands is not likely to occur.

What is the present situation?

The present situation is that much remains to be accomplished to bring self-sufficiency to the Third World. Many individual nations, such as Mexico, are no longer self-sufficient. Africa has suffered declining food production since 1960.

CHAPTER 25—SAMPLE OBJECTIVE TEST QUESTIONS

A. MATCHING

1.	consumers	a. _______	all living organisms
2.	producers	b. _______	animals
3.	decomposers	c. _______	energy flow in steps
4.	biomass	d. _______	green plants
5.	macronutrient	e. _______	light energy
6.	food chain	f. _______	oxidation
7.	photosynthesis	g. _______	bacteria
8.	respiration	h. _______	weight of organic matter
9.	biosphere	i. _______	carbon

B. MULTIPLE CHOICE

1. Photosynthesis requires each of these three ingredients:

   ________a. nitrogen, oxygen, light energy.

   ________b. carbohydrate, water, chemical energy.

   ________c. water, carbon dioxide, light energy.

   ________d. water, carbohydrate, light energy.

2. Net primary production is

   ________a. high over deep oceans, far from land.

   ________b. high in deserts.

   ________c. low in upwelling zones.

   ________d. high in wet equatorial climate.

3. Agricultural ecosystems

   ________a. require large inputs of fertilizers.

   ________b. are efficient in recycling of plant nutrients.

   ________c. require many genetic strains of a given crop in the same field.

   ________d. are highly resistant to insect destruction.

C. COMPLETION

1. A metabolic process in which organic compounds are oxidized within living cells is called ______________________________ in the process. ______________________________ is lost to the outside.

2. Gross photosynthesis minus respiration is called ______________________________; it can be used as a measure of the intensity of ______________________________.

3. Energy flow upward in the food chain results in an energy loss of from _______________ % to _______________% in each step.

4. The nine elements required in greatest abundance for organic growth are termed___________ _______________________; the top three in order of abundance, are.______________________, _______________________, and _______________________________.

CHAPTER 26

CONCEPTS OF BIOGEOGRAPHY

PAGE

What is *biogeography?*

Biogeography: study of the distribution patterns of plants and animals on the earth's surface and the processes that produce those patterns.

What is the scale range on which the physical environment of plants and animals can be treated?

Scale range on which environmental influences on plants and animals are treated: (1) global scale, considering climatic factors; (2) local, small-area variations (habitats, communities).

What are the most important factors governing plant and animal distribution on all scales?

Air temperatures and soil-water availability are the major factors governing plant and animal distribution at both global and local scales.

BIOLOGICAL ROLE OF WATER

BIOLOGICAL ROLE OF WATER 502

How important is water in determining global distribution patterns of plants and animals?

Water is most important of the factors that determine global distribution patterns of plants and animals, since these organisms have become specialized and adapted to degree of water availability.

How do plants affect the water balance near the earth's surface?

Plants, by transpiration, return water to atmosphere from soil, thus altering water balance of the surface layer.

ORGANISMS AND WATER NEEDS

ORGANISMS AND WATER NEEDS 502

How are plants classified by water need? What is the meaning of these prefixes: *xero, hygro, meso?*

Plants classified by water need: Prefixes *xero,* dry, *hygro,* wet; *meso,* intermediate, used in three classes:

Define *xerophyte.*

Xerophyte: plant adapted to dry environment.

Define *hygrophyte.*

Hygrophyte: plant adapted to wet environment.

Define *mesophyte.*

Mesophyte: plant adapted to habitat of intermediate degree of wetness and uniform soil water availability.

What factors cause a high rate of transpiration from plant foliage?

High rate of transpiration from plant foliage favored by high temperatures, low humidities, and high winds.

How are plant leaves adapted to control transpiration rates?

Transpiration rates controlled by structures of plant leaves by means of specialized pores, called *stomata.*

What are *stomata?*

Stomata: specialized leaf pores that are openings in the outer cell layer, or *cuticle,* through which transpiration occurs.

What is the *cuticle?*

Cuticle: outermost protective cell layer of a leaf.

How do the stomata control transpiration?

Stomata are surrounded by *guard cells* that can open and close to regulate transpiration of water vapor and outflow of other gases.

What other structures or materials on plant leaves serve to reduce water losses?

Water loss is reduced in some plants by thickening of cuticle and deposition of wax coatings, by deeply sunken stomata, by reduced leaf area, or by having no leaves.

How do plant roots respond to a water-scarce environment.

Plant roots in water-scarce environment can be greatly extended downward, may reach ground-water table as steady source, or roots may be widespread but shallow to absorb water from sporadic downpours.

What are *phreatophytes?* Where are they found?

Phreatophytes: plants with deep roots that draw water from the ground-water table beneath alluvium of dry stream channels and valley floors in desert regions.

How are stems of desert plants adapted to store large amounts of water? What are *succulents?*

Succulents: plants adapted to resist water loss by means of thickened spongy tissue in which water is stored. Example: cactus.

How do *ephemeral annuals* show an adaptation to extreme aridity of the environment?

Ephemeral annuals: small desert plants that complete a life cycle very rapidly following a desert downpour.

How are plants adapted to a wet-dry tropical climate or a cold climate in which water is unavailable in the winter season? What is a *tropophyte?*

Tropophyte: plant that sheds its leaves and enters a dormant state during a dry or cold season when soil water becomes unavailable.

What is a *deciduous plant?* Is a deciduous plant a tropophyte?

Deciduous plant: tree or shrub that seasonally sheds its leaves, i.e., a tropophyte.

What is an *evergreen plant?*

Evergreen plant: tree or shrub that holds most of its green leaves throughout the year.

What are *sclerophylls?*

Sclerophylls: hard-leaved evergreen trees and shrubs capable of enduring a long dry summer.

Under what soil-water regime and climate are sclerophylls found?

Sclerophylls are typical of the xeric soil-water regime of the Mediterranean climate (7); has long dry summer, wet winter.

What are *xeric animals?*

Xeric animals: animals adapted to dry conditions typical of a desert climate.

Give examples of ways in which xeric animals cope with a shortage of water.

Xeric animals that are invertebrates may remain in dormant stages in dry periods. Example: brine shrimp complete life cycles in brief period when desert lake basins hold water.

How do desert mammals avoid excessive water loss?

Desert mammals conserve water by reduced sweating, by excreting concentrated urine and dry feces, by nocturnal habit that avoids high air temperatures.

ORGANISMS AND TEMPERATURE

ORGANISMS AND TEMPERATURE 504

In what ways does air and soil temperature influence plant functions? Name those functions.

Air and soil temperture strongly influences plant functions of photosynthesis, flowering fruiting, seed germination, and leaf shedding.

How does air temperature act indirectly upon plants? What functions are affected?

Air temperature acts indirectly upon plants by regulating water-vapor capacity of air; affects transpiration and evaporation of soil water.

How does cold act upon plants? How does a cold climate affect numbers of species? What effect has freezing upon plant tissues?

Cold limits numbers of specied capable of survival. Few species exist in cold subarctic and arctic environment. Freezing disrupts plant tissues holding water.

How do animals moderate the effects of temperature variations?

Animals moderate effects of temperature variations by their physiology and by ability to seek sheltered environments.

What are *cold-blooded animals?*

Cold blooded-animals: animals whose body temperature passively follows the temperature of the environment: reptiles, invertebrates, fishes, amphibians.

What happens to the cold-blooded animals during the winter season?

Cold-blooded animals (fishes excepted) become dormant in winter: vertebrates may enter state of *hibernation.*

What is *hibernation?*

Hibernation: dormant state of some vertebrates in winter season during which metabolic porcesses nearly halt and body temperatures closely parallel those of surroundings. Most hibernators seek burrows where winter soil temperature is moderate.

What are *warm-blooded animals?*

Warm-blooded animals: animals that possess one or more adaptations to maintain a constant internal temperature despite fluctuations in the environmental temperature.

Give examples of the adaptations used by warm-blooded animals.

Warm-blooded animals may have fur, hair, or feathers to insulate body, or a thick fat layer. Cooling achieved by sweating or panting, by exposed blood-circulating tissues.

OTHER CLIMATIC FACTORS 505

Explain how the factor of light influences forest plants.

Factor of light important in determining local plant distribution patterns. Tree crowns in forest receive maximum; cut off light from lower layers and forest floor. Low annual plants may flourish in spring, before tree canopy is completed. Plants requiring deep shade appear later.

How does duration of daylight influence global plant distribution?

Duration of daylight in summer increases with higher latitude, becomes 24 hrs at arctic circle. Although growing season is short in high latitudes, prolonged daylight greatly accelerates plant growth.

How does the annual cycle of change of the daylight cycle influence plants in midlatitudes?

Annual cycle of daylight in midlatitudes governs timing of budding, flowering, fruiting, and leaf-shedding.

How does light influence animal behavior?

Animal activities are controlled by day-night cycle; some are active in daylight, others at night. Seasonal animal activities controlled by *photoperiod* in midlatitudes; examples, food gathering, mating, and reproduction.

What is *photoperiod?*

Photoperiod: duration of daylight on a given day at a given latitude.

How does wind influence plants?

Wind is important environmental factor for plants; deforms tree shapes in arctic and alpine zones, affects elevation of tree limit.

BIOCLIMATIC FRONTIERS 506

What is a *bioclimatic frontier?*

Bioclimatic frontier: a geographic boundary corresponding with a critical limiting level of climatic stress beyond which a species cannot survive.

Give an example of a bioclimatic fronties.

Example: Ponderosa pine in western N. America grows largely within the 50 cm (20 in.) isohyet of precipitation.

Can plants or animals be held within geographic limits by factors other than climate?

Geographic limits to plant or animal species may also be controlled by diseases, predation, dependency on other species, or slow rates of outward migration.

INTERACTIONS AMONG SPECIES 506

Name the three ways two species can interact as part of the same ecosystem.

The three ways two species can interact as part of the same ecosystem are: (1) interaction may be negative to one or both species, such as *competition, parasitism, predation, herbivory and allelopathy;* (2) interaction may be neutral; and (3) interaction may be positive benefiting one of both species, such as commensalism, protocooperation, and mutualism.

What is *competition* between two species?

Competition: when two species require a common resource that is in short supply.

What is the result of this competition?

The result of this competition: sometimes one species will win and crowd out competitor. At other times they may compete indefinitely.

What is *predation* and *parisitism?*

Predation and *parisitism:* negative interactions in which one species feeds on the other. In *predation*, the organism which gains energy is larger; in *parisitism*, the organism gaining energy is smaller.

What is *herbivory?*

Herbivory: grazing of plants by animals which reduces the viability of the plant population.

Define *allelopathy:*

Allelopathy: chemical toxins produced by one species inhibits the growth of others.

What three positive interactions are included in *symbiosis?*

The three positive interactions included in symbiosis are (1) commensalism: one species is benefited and the other is unaffected; (2) protocooperation: the relationship benefits both parties, but is not essential to their existence; (3) mutualism: protocooperation progresses to the point where one or both species cannot live alone.

TERRESTRIAL ECOSYSTEMS—THE BIOMES

TERRESTRIAL ECOSYSTEMS—THE BIOMES 508

What are the two major groups of ecosystems?

Two major groups of ecosystems: (1) *aquatic*; (2) *terrestrial*.

What are *aquatic ecosystems*?

Aquatic ecosystems: ecosystems consisting of life forms of the marine environments and the freshwater environments of the lands.

What are the *terrestrial ecosystems?* Where are they found?

Terrestrial ecosystems: plant and animal communities occupying land surfaces of the continents. (Excludes bogs, swamps, streams, and ponds.) Terrestrial ecosystems are directly impacted by climate and interact with soil.

What is a *biome?* What organisms are included within a biome? What kind of organism makes up most of the biomass of a biome?

Biome: the largest unit of terrestrial ecosystem; the total assemblage of plant and animal life associated with a major life-form unit controlled by climate. Biomass of green plants makes up the bulk of a biome.

Naame the five principal biomes. For each, give conditions of availability of soil water and heat.

Biomes recognized through availability of soil water and heat:

Forest (ample soil water and heat)

Savanna (transitional between forest and grassland)

Grassland (moderate soil-water shortage; adequate heat)

Desert (extreme soil-water shortage; adequate heat)

Tundra (insufficient heat)

Define *formation classes.* On what basis are formation classes set apart from one another? With what geographical factors can the formation classes be correlated?

Formation classes: vegetation units based upon life-form and comprising subdivisions of the biomes. Formation classes can be closely correlated with major types of climates, water budgets, and soils.

COMMUNITIES AND HABITATS

COMMUNITIES AND HABITATS 508

What are *biotic communities?*

Biotic communities: local associations of plants and animals that are interdependent and often found together.

What factors cause place-to-place variations among communities?

Primary influences on communities are landform (upland, valley) and soil (well-drained, poorly drained).

Define *habitat.* What physical factors make up the habitat?

Habitat: a subdivision of the plant environment having a certain combination of slope, drainage, soil type, and other controlling physical factors.

Name several typical forest habitats. Which habitat is usually regarded as typical for description of plant life-form classes?

Typical forest habitats: upland, bog, bottomland, ridge, cliff, active dune. Upland habitat usually regarded as typical for description of plant life-form classes.

GEOMORPHIC FACTORS

GEOMORPHIC FACTORS 509

What geomorphic factors influence ecosystems?

Geomorphic factors (landform factors) influencing ecosystems: slope steepness, slope aspect, relief. Geomorphic factors include character of landforms as shaped by erosion, transportation and deposition by streams, waves, wind and ice.

How does slope steepness influence soils and plants?

Slope steepness influences rate of drainage and infiltration of precipitation; steep slopes, rapid drainage and rapid erosion; gently slopes, slow drainage, slow erosion, thick soil.

How does slope aspect influence soils and plants?

Slope aspect affects exposure to sunlight and prevailing winds. Slopes facing sun are drier, warmer; slopes in shade are cooler, moister. Plant communities respond.

EDAPHIC FACTORS

EDAPHIC FACTORS 509

What are *edaphic factors?*

Edaphic factors: factors influencing a terrestrial ecosystem that are related to the soil.

What two viewpoints can be taken on assessing edaphic factors?

Two viewpoints on edaphic factors: (1) global patterns of soils as controlled by climate; (2) small-scale place-to-place variations in soil are a result of landform controls.

Name some edaphic factors that influence plant habitats and communities.

Examples of edaphic factors influencing plant habitats and communities: soil texture and structure; humus content; alkalinity, acidity, or salinity; biological activity in soil.

In what way are plants directly involved in the evolution of the soil?

Given a barren soil parent material, soil evolution goes hand in hand with occupation of the habitat by organisms. Plants and animals contribute to soil formation.

ECOLOGICAL SUCCESSION

ECOLOGICAL SUCCESSION 510

Define *ecological succession.* Where does succession occur?

Ecological succession: a time-succession of distinctive plant and animal communities occurring within a given area of newly formed land, or land burned over, or otherwise cleared of plant cover.

What is the general trend of ecological succession?

Trend of ecological succession leads to formation of the most complex community of organisms possible in a given area.

What is a *sere?*

Sere: in an ecological succession, the series of biotic communities that follow one another on the way to the stable stage, or *climax.*

What are *seral stages?*

Seral stages: stages within a sere.

What is a *climax?*

Climax: stable biotic community of plants and animals reached at the end of a sere.

What two kinds of succession are recognized?

Two kinds of ecological succession *primary succession, secondary succession.*

Define *primary succession.*

Primary succession: ecological succession that begins on a newly constructed deposit of mineral sediment.

Define *secondary succession.*

Secondary succession: ecological succession beginning on a previously vegetated area that has been recently disturbed by such agents as fire, flood, windstorm, or human activities.

Give examples of the parent soil materials on which primary succession may begin.

Materials on which primary succession may begin: sand dunes, sand beaches, fresh lava flows, layers of volcanic ash, new silt deposits on inside of stream bend.

How does the natural fertility of the parent material vary with type of deposit?

Some primary materials (sand dunes, beach sand) lack colloids and nutrient bases; others (alluvial silts) may be well endowed with colloids and exchangeable base cations.

Describe the *pioneer stage* in succession.

Pioneer stage: first stage in a succession. Consists of a few species adapted to adverse environmental conditions, such as rapid water drainage, exposure to intense sunlight and wind, extreme ground temperatures.

How does plant growth lead to environmental changes?

Plant growth adds humus to soil layer; adds surface accumulation or organic matter; provides shade.

What happens as succession continues?

As environmental conditions change, other species invade the area, displacing pioneers. Plant cover becomes denser, changing the *microclimate.*

What is *microclimate?*

Microclimate: climate of a shallow layer of air near the ground and including the soil surface and plant community within which it is in contact.

How does microclimate change during succession?

As succession continues, microclimate shows less extreme air and soil temperatures, higher humidity, because plant cover becomes denser.

Describe primary succession on coastal sand dunes in a moist climate.

Primary succession on coastal dunes: pioneer stage has beachgrass; low woody shrubs follow, then larger woody plants and trees. Climax may consist of broadleaf forest. Animal life also changes in composition during succession.

Describe *bog succession.*

Bog succession: primary succession occurring on a shallow lake under cold continental climate in areas recently glaciated. Aquatic plants (rushes, sedges) begin succession, followed by sphagnum (peat moss), filling lake; peat deposit supports hygrophytic forest trees.

What is *freshwater peat?*

Freshwater peat forms of peat produced by bog succession, often identified as the organic soil horizon of a Histosol.

OLD-FIELD SUCCESSION

OLD-FIELD SUCCESSION 512

What is *old-field succession?*

Old-field succession: form of secondary succession typical of an abandoned field such as might be found in the moist continental climate or the moist subtropical climate of the eastern and central Unites States; a form of *autogenic succession.*

What is an *autogenic succession?*

Autogenic succession: form of ecological succession that is self-producing i.e., the result of the action of the plants and animals themselves.

Describe briefly the stages in old-field succession.

Stages in old-field succession: pioneers are annuals and biennials (weeds) or perennial herbs and grasses; pine forest follows, to be replaced by deciduous hardwood forest (oak, hickory) as climax.

HUMAN IMPACT ON NATURAL ECOSYSTEMS

In what ways, besides intensive agriculture and urbanization, does human influence natural vegetation?

HUMAN IMPACT ON NATURAL ECOSYSTEMS 512

Humans influence natural vegetation by preventing natural forest fires, by transplanting species from one continent to another, by introduction of plant diseases and predatory insects.

THE GREAT YELLOWSTONE FIRE—DISASTROUS OR BENEFICIAL? 513

What is the general setting for Yellowstone National Park?

Yellowstone National Park occupies about 900,000 hectares (2.2 million acres) in northwestern Wyoming. It contains a collapsed caldera, so volcanic rocks and geothermal activity in the form of hot springs and geysers occur. Its forests and parks support a rich population of wildlife.

Describe the fires that occurred there.

In August and September of 1988, many fires burned out of control in the Park. Most were started by lightening strikes. Strong winds fanned the flames. Because the summer had been the driest for more than a century, the fires burned vigorously and consumed many stands of older trees. Even some young stands burned.

What was the result?

Only 20 percent of the Park was burned to some degree. Most of the large animals survived the fire and the loss of forest. Mammals, insects and birds returned quickly to the burned areas.

Why was the Park Service criticized?

The Park Service was criticized for letting the fires, which started in July, burn naturally at first. This was in accordance with a new policy that fires were natural and should not be contained. However, a massive effort in late July failed to put out the established fires.

What did forest ecologists conclude about the fires?

Forest ecologists concluded that, once started, little could have been done to stop the fires. The "no-burn" policy of suppressing fires immediately, which had been followed for nearly one hundred years, only enhanced the fire spread by a small amount.

Describe the natural fire cycle in Yellowstone.

There is a natural cycle of great fires about every 250 to 300 years. After such a fire, lodgepole pine trees come in and grow densely for 150 to 250 years. After this time, the pine canopy can burn, restarting the cycle. If the pines do not burn, they die and are replaced by spruce and fir. However, the dead pines on the ground make the spruce-fir forest vulnerable to burning as well.

How does this cycle appear on the landscape?

Because the fires burn patches instead of large areas, the forest is a mosaic of patches in different stages of succession. The young patches are less flammable, and provide natural firebreaks, limiting the spread of the fire and sheltering the animals. After the fire, the new ground cover of greasses and shrubs provides needed food for grazing animals.

BIOMASS BURNING AND ITS IMPACTS ON THE ATMOSPHERE 514

What is biomass burning? Where is it prevalent?

Biomass burning: combustion of plant matter in natura and human-set fires. It is prevalent in the low latitude zones.

What is burned and why?

Trees and shrubs are burned in clearing rainforest for agriculture. Dead grasses and seedling trees and shrubs are burned to promote a new crop of dense grasses for grazing. Agricultural waste products are burned for disposal. Wood is burned for cooking fires. Discarded wood and wood products are burned for disposal.

How much carbon is released into the atmosphere per year by burning? What gases are released?

Between 2 and 5 billion metric tons of carbon is released to the atmosphere by burning. CO_2 gas is the major product; others are methane, forms of nitrogen gases, hydrogen, hydrocarbons, and sulfur dioxide.

What is the impact of smoke?

Rising smoke plumes can stimulate rain in some cases. Some smoke particles reach the stratosphere, where they can exert a cooling effect on global temperature. Smoke constituents can induce production of ozone air pollution near the ground and contribute to acid deposition in rainfall.

What is the contribution of biomass burning to global climate?

Carbon dioxide from biomass burning may be responsible for as much as 20 to 60 percent of predicted greenhouse warming. Methane and low-altitude smoke also contribute. Smoke particles may also promote cloud formation, which could favor either warming or cooling, depending on the altitude and extent of the cloud cover.

CHAPTER 26—SAMPLE OBJECTIVE TEST QUESTIONS

A. MATCHING

1.	hygrophyte	a. _______	dry
2.	phreatophyte	b. _______	hard leaf
3.	xeric	c. _______	climax
4.	stomata	d. _______	bog
5.	sclerophyll	e. _______	wet
6.	sere	f. _______	guard cells
7.	peat	g. _______	ground water

B. MULTIPLE CHOICE

1. A tropophyte is a plant that

   ________a. sheds its leaves during one season of year.

   ________b. sheds its leaves during the summer.

   ________c. resists water loss by developing thorns.

   ________d. is a form of hygrophyte.

2. Edaphic factors are those factors in biogeography related to

   ________a. landform.

   ________b. plant habitat.

   ________c. soil.

   ________d. plant succession.

3. In the climax stage of old-field succession

   ________a. pioneers predominate.

   ________b. annual herbs predominate.

   ________c. pine forest predominates.

   ________d. a forest of oak and hickory is present.

C. COMPLETION

1. Plants that grow in dry habitats are called ____.
2. Surrounding the stomata of a plant leaf are ____ cells; they serve to regulate ____.
3. Plants that appear briefly on the desert floor after a heavy downpour are ________________ ____ annuals.

4. A dormant state assumed by animals during the cold season is ______________________________ ______________________________.

5. Length of daylight period is known as the ______________________________.

6. A geographic species boundary determined by climatic factors is the ______________________________ ______________________________ frontier.

CHAPTER 27

WORLD PATTERNS OF NATURAL VEGETATION

PAGE

DESCRIBING THE STRUCTURE OF VEGETATION — 516

What categories comprise the structural description of vegetation?

Structural description of vegetation is based on physical properties and outward form of plants. Six categories of information (Dansereau system): (1) life-form; (2) size and stratification; (3) coverage; (4) periodicity; (5) leaf shape and size; (6) leaf texture.

What is meant by *life-form* of plants?

Life-form: classification of plants in terms of outward form, including *tree, shrub, liana, herb, bryoid.*

Define *tree.* Describe the structure of a tree.

Tree: large, erect, woody, perennial plant with single main trunk, few branches in lower part, and a branching crown.

Define *shrub.* How does a shrub differ from a tree?

Shrub: woody perennial plant, usually small or low, with several low-branching stems and a foliage mass close to the ground.

Define *liana.*

Liana: woody vine supported on trees.

Define *herb.* Are both annuals and perennials included in herbs?

Herb: tender plant, lacking woody stems, usually small or low. Herbs include annual and perennial plants. Some herbs are *forbs;* others are grasses.

What is a *forb?*

Forb: broadleaved herb, as distinguished from grasses.

What is meant by *herbaceous?*

Herbaceous: adjective applied to plants that are herbs.

What are *bryoids?*

Bryoids: group of low, small plants that lie close to the ground or are attached to tree trunks; most are mosses and liverworts.

What are *epiphytes?*

Epiphytes: plants that live above ground level out of contact with the soil, usually growing on the limbs of trees or shrubs; also called air plants.

What are *thallophytes?*

Thallophytes: plants lacking true roots, stems and leaves, including: bacteria, algae, molds and fungi.

What are *lichens?*

Lichens: plant forms in which algae and fungi live together in a symbiotic relationship to form a single structure; they typically form tough, leathery coating or crusts attached to rocks and tree trunks.

How is plant size described?

Plant size: is defined in terms of height range in assigned values for each life-form class (example: low tree, 8 to 10 m).

How is *coverage* of vegetation described?

Coverage: degree to which foliage of plants of a given life-form covers the ground. Terms used: barren or very sparse, discontinuous, in tufts or groups, continuous.

How is periodicity described?

Periodicity: response of plant foliage to the annual climatic cycle. Plants are classed as (1) *deciduous;* (2) *evergreen;* (3) *semideciduous*; and (4) *evergreen-succulent* or *evergreen-leafless.*

Define *deciduous plant.*

Deciduous plant: tree or shrub that sheds its leaves seasonally, i.e., a tropophyte.

Define *evergreen plant.*

Evergreen plant: tree or shrub that holds most of its green leaves throughout the year.

Define *semideciduous plant.*

Semideciduous plant: plant that sheds its leaves at intervals not in phase with a climatic season.

Define *evergreen-succulent plants.*

Evergreen-succulent plant: evergreen plant with thick, fleshy leaves or stems that retain foliage year around.

Define *evergreen-leafless plant.*

Evergreen-leafless plant: evergreen plant with fleshy stems but no functional leaves. Example: cacti.

How are leaf shape and size described?

Leaf shape and size: leaf forms include *broadleaf, needleleaf, spine, graminoid, small leaf,* and *compound leaf.*

Define *broadleaf.*

Broadleaf: leaf form that is wide in relation to length, thin, and comparatively large. Examples: sycamore, maple.

Define *needleleaf.*

Needleleaf: leaf form that is very narrow in relation to length. Examples: spruce, fir, pine.

Define *spine.*

Spine: leaf form that is a hard, sharp-pointed spike.

Define *graminoid.*

Graminoid: plant form having long, narrow leaves; i.e., grasslike. Example: rye grass.

Define *small leaf.*

Small leaf: leaf form in which leaves are thin and comparatively wide, but of small overall dimensions as compared with the broadleaf form. Examples: birch, holly.

How is leaf texture described?

Leaf texture: quality of thickness or hardness of the leaves of a plant species. Textures include *membranous, filmy, sclerophyllous,* and *succulent.*

What are *membranous* leaves?

Membranous leaves: leaves of the broad leaf type that are of normal thickness.

What are *filmy leaves?*

Filmy leaves: leaves that are thin and delicate as compared with membranous leaves. Example: maidenhair fern.

Describe *sclerphyllous* leaves.

Sclerophyllous leaves: leaves of sclerophylls; they are hard, thick, and leathery. Example: eucalyptus.

Define *sclerophyll forest.*

Sclerophyll forest: forest dominated by trees and shrubs having sclerophyllous leaves.

Describe *succulent* leaves.

Succulent leaves: leaves that are very greatly thickened, with spongy structure, capable of holding much water.

MAJOR BIOMES

MAJOR BIOMES 519

Review the definition of a biome.

Biomes: broad major groups of natural ecosystems that include both animal and plant life. Major biomes are *forest, savanna, grassland, desert,* and *tundra.*

What is the *forest biome?*

Forest biome: biome consisting of all regions of *forest.*

Define *forest.*

Forest: assemblage of trees growing close together, their crowns forming a layer of foliage that largely shades the ground. Forest may show stratification, with more than one layer.

What are the water requirements of a forest? With what climate types is forest associated?

Forests require relatively large annual precipitation. Soil-water balance is most closely associated with humid and perhumid subtypes of the moist climates (Chapter 9). Soil-water storage remains high through most of the year. Forest biome ranges in latitude from wet equatorial climate (1) to boreal forest climate (11).

Describe the *savanna biome.*

Savanna biome: biome consisting of regions of combination of trees and grassland in various proportions. Parklike in appearance; trees spaced singly or in groups, interspersed with surfaces covered by grasses or other plant forms of low height.

With what climate is the savanna biome associated?

Savanna biome is associated with warm climates with alternate wet and dry seasons: wet-dry tropical climate (3) and semiarid subtype of dry tropical climate (4s).

Describe the *grassland biome.*

Grassland biome: biome consisting of regions largely or entirely formed of herbs, which may include grasses, grasslike plants, and forbs.

With what climates is the grassland biome associated?

Grassland biome is associated with semiarid subtype of dry midlatitude climate (9s) and with subhumid subtypes of the moist subtropical climate (6sh) and moist continental climate (10sh). Soil-water shortage 0–20 cm or larger; no water surplus.

Describe the *desert biome.*

Desert biome: biome consisting of regions of extremely arid climates with thinly dispersed plants. Much of the ground surface is bare. Although treeless, desert biome may have woody plants, grasses, and perennial herbs.

Describe the *tundra biome.*

Tundra biome: cold-climate biome consisting of grasses, grasslike plants, flowering herbs, dwarf shrubs, mosses, and lichens. Found both in arctic zone (arctic tundra) and at high altitudes (alpine tundra).

FORMATION CLASSES

FORMATION CLASSES 522

What are formation classes?

Formation classes: subdivisions within the biome, bases on the size, shape, and structure of the plants that dominate the vegetation.

Do formation classes relate to climate types and soil classes?

Formation classes have close relationships with climate types and subtypes; with soil orders and suborders.

Does species composition (floristic composition) enter into the definition of formation classes?

Species composition (floristic composition) is basis for recognition of certain formation classes.

FOREST BIOME

FOREST BIOME 522

Equatorial and Tropical Rainforests

Equatorial and Tropical Rainforests 522

Describe the *equatorial rainforest.*

Equatorial rainforest: forest of tall, closely set trees with crowns forming a continuous canopy.

Describe the trees of the equatorial rainforest.

Trees are smooth-barked, unbranched in lower two-thirds, often buttressed at base.

Describe the leaf canopy of the equatorial rainforest trees.

Trees are evergreen, broadleaved. Crowns form two or three layers with scattered emergent crowns.

Describe lianas of the equatorial rainforest. Are epiphytes numerous?

Lianas of rainforest are thick and woody, rise to upper level and may have large crowns. Epiphytes are numerous; some are *stranglers.*

What are *stranglers?*

Strangler: twining plant (liana) that surrounds a living tree trunk, eventually killing and replacing that tree.

What can be said about the numbers of tree species in the equatorial rainforest?

Numbers of different species of equatorial rainforest is very great within a given small area.

Describe the floor of the equatorial rainforest.

Floor of equatorial rainforest is densely shaded, may be open in aspect. Leaf litter is thin, humus lacking, because of rapid rate of consumption of dead plant matter.

With what climate is the equatorial rainforest associated geographically?

Equatorial rainforest closely associated with wet equatorial climate (1) and monsoon and trade-wind littoral climate (2).

Describe the soil-water balance of the equatorial rainforest.

Soil-water balance has large water surplus; high level of soil-water storage in all months.

With what soil orders is the equatorial rainforest associated?

Oxisols and Ultisols are soil orders closely associated with rainforest. Nutrient supply is meager; mostly in shallow surface layer and stored in biomass of forest.

What formation class is found in disturbed areas of rainforest?

Disturbed areas may have *jungle* formation.

What is *jungle*?

Jungle: low, dense, nearly impenetrable plant formation consisting of lianas, bamboo, scrub, thorny palms, or thickly branching shrubs.

What coastal vegetation type is found along low-lying coasts in the low latitudes?

Coastal vegetation along shallow, muddy zones is *mangrove swamp forest.*

Describe *mangrove swamp forest.*

Mangrove swamp forest: coastal forest of stilted mangrove trees with prop roots that trap muddy sediment.

How does *tropical rainforest* differ from equatorial rainforest?

Tropical rainforest: forest formation class similar in most respects to equatorial rainforest, but with fewer species and fewer lianas.

Describe the *montane forest.* Where does it occur?

Montane forest: high-altitude equatorial and tropical zone forest on mountain ranges with strong orographic effect. Trees are lower than in rainforests. Tree ferns, bamboo, epiphytes, are numerous. Heavy moss accumulations on limbs at higher elevation in *mossy forest.*

What is *mossy forest?*

Mossy forest: high-altitude forest in low latitudes; characterized by massive accumulations of mosses on tree limbs.

What is *elfin forest?*

Near upper limit of forest is *elfin forest,* of moss-festooned stunted trees.

EXPLOITATION OF THE LOW-LATITUDE RAINFOREST ECOSYSTEM 525

Describe subsistence agriculture in the rainforest.

Slash-and-burn: shifting agricultural system practiced in low-latitude rainforest, in which small areas are cleared and burned, forming plots that can be cultivated for brief periods.

How large an area of low-latitude rainforest is removed annually?

In 1989, the global total of low-latitude rainforest removed was estimated at about 16 to 20 million hectares. This is an area about the size of the state of Washington.

What countries experienced significant deforestation?

Countries experiencing significant deforestation included Brazil, India, Indonesia, Burma, Cameroon, Costa Rica, the Philippines, and Vietnam.

What is the potential effect of deforestation on local climate?

After deforestation, soil and air temperatures can increase, and evaporation and precipitation decrease. Dry seasons are extended.

What organizations are working to protect rainforests?

Organizations working to protect rainforests include the United Nations Food and Agricuture Organization and the International Tropical Timber Organization.

What is happening to rainforest in Sarawak?

Logging of rainforest in Sarawak is not being conducted in a sustainable manner, and all of Sarawak's primary rainforests will be destroyed in a decade if current trends continue.

How can the rainforest be protected?

One way of protecting the rainforest is to protect and strengthen the economic base of the native inhabitants whose use of forest resources fits the environment. This includes harvesting of food products and medicinal plants from the rainforest.

Monsoon Forest

Monsoon Forest **527**

Describe *monsoon forest.*

Monsoon forest: tropical zone forest formation class characterized by deciduous trees that shed leaves in cool, dry season (time of low sun).

With what climate and soil types is monsoon forest associated?

Monsoon forest is associated with wet-dry tropical climate (3). Soils are largely Ultisols and Oxisols on upland areas; also some areas of Ustalfs.

Broadleaf Evergreen Forest

Broadleaf Evergreen Forest **528**

Describe *broadleaf evergreen forest.*

Broadleaf evergreen forest (also called *temperate rainforest*, or *laurel forest*): forest formation class dominated by broadleaf evergreen trees. Species are few in umber compared with low-latitude rainforests; leaves smaller and more leathery, tree canopy less dense.

What species of trees are common in the broadleaf evergreen forest?

Common species types: SE United States, evergreen oak, laurel, magnolia; New Zealand, tree ferns, kauri, podocarps, small-leaved beeches.

With what climate and soil types is the broadleaf evergreen forest associated?

Climate is typically moist subtropical (6) or marine west-coast (8). Soils are typically Udults (order Ultisols) with low nutrient base content in lower profile; also Udalfs (Australia).

Midlatitude Deciduous Forest

Midlatitude Deciduous Forest **530**

Describe the *midlatitude deciduous forest.*

Midlatitude deciduous forest: forest formation class dominated by tall, broadleaf deciduous trees. (Also called *summergreen deciduous forest.*)

Name some common tree species of the midlatitude deciduous forest in U. S. and Europe.

Common tree species: U. S. oak, beech, birch, hickory, walnut, maple, basswood, elm, ash, tulip, chestnut, hornbeam; Europe: oak, ash, beech.

With what climate and soil types is the midlatitude deciduous forest associated?

Climate is moist continental (10h, 10p) in eastern N. America and Europe; marine west-coast (8h) in British Isles, W. Europe. Soils are typically Udalfs and Boralfs (order Alfisols).

Needleleaf Forests

Needleleaf Forests 530

Describe *needleleaf forest.*

Needleleaf forest: forest formation class consisting largely of straight-trunked, conical, needleleaf trees that are conifers. Species are few.

What subtypes are found within the needleleaf formation class?

Subtypes within needleleaf forest: *boreal forest, needleleaf evergreen forest, coastal needleleaf evergreen forest, lake forest, southern pine forest* and *cold woodland.*

What is *boreal forest?* Where does it occur?

Boreal forest: needleleaf forest of the boreal forest climate (11) in North America and Eurasia. Consists of evergreen conifers (spruce, fir, pine) in N. American, Europe, and western Siberia; of deciduous larch in central and eastern Siberia, where climate shows semiarid soil-water balance. Soils are Spodosols, Boralfs, Histosols, Cryaquepts.

What is *needleleaf evergreen forest?* Where does it occur?

Needleleaf evergreen forest: found at high altitudes in mountain ranges and plateaus of midlatitude zone.

What is *coastal needleleaf evergreen forest?* Where does it occur?

Coastal needleleaf evergreen forest: found in coastal ranges of British Columbia, California; consists of dense coniferous trees such as redwood, big tree, Douglas fir.

What is *lake forest?*

Lake forest: type of needleleaf evergreen forest formerly in Great Lakes Region; was made up of white and red pine, hemlock. Has been largely destroyed by lumbering.

What is *cold woodland?* With what climate is it associated?

Cold woodland (taiga): woodland of scattered needleleaf trees, with open areas of lichens and mosses. Associated with northern fringe of boreal forest climate (11), transitional to tundra.

What is *muskeg?* Where is it found?

Bog areas of glaciated N. America and Europe have hygrophytic forest growing on *muskeg,* a thick organic (peat) layer filling former glacial lakes.

Sclerophyll Forest

Sclerophyll Forest 532

Describe *sclerophyll forest.*

Sclerophyll forest: formation class consisting of low trees with small, hard, leathery leaves (sclerophyllous leaves). Trees are typically low-branched, gnarled, with rough bark. Formation class includes sclerophyllous *woodland* and *shrub.*

What is *woodland?*

Woodland: open forest with canopy coverage 25 to 60%.

What is *scrub?*

Scrub: formation type consisting of shrubs with canopy cover about 50%.

With what climate is sclerophyll forest associated?

Climate of sclerophyll forest regions is Mediterranean type (7), along west coasts in latitude range 30°–45°.

Describe the sclerophyll forest of Mediterranean lands (*Mediterranean evergreen mixed forest*).

In Mediterranean lands, sclerophyll forest is an evergreen mixed forest, consisting of cork oak, live oak, Aleppo pine, stone pine, and olive. Most has been destroyed, replaced by *maquis*, a dense scrub.

Describe the sclerophyll forest and woodland of California.

Sclerophyll forest and woodland of California coast ranges consists of live oak and white oak, or of sclerophyllous scrub, known as *chaparral*. Chaparral may consist of wild lilac, manzanita, mountain mahogany, poison oak, live oak.

Describe the *sclerophyll forest of Australia*.

Australian sclerophyll forest consists of *Eucalyptus* and acacia.

With what soils are sclerophyll forest, woodland, and scrub associated?

Sclerophyll vegetation is associated with Xeralfs, Xerolls, Xerults; all of xeric soil-water regime.

FORESTS AND GLOBAL WARMING

How do young and old forests compare in their withdrawl of CO_2 from the air? What might we conclude from this?

A young forest withdraws CO_2 from the atmosphere and stores the carbon within its biomass. Very old forests increase their biomass very little, so they withdraw little CO_2. From this, we might conclude that cutting down old forests and replacing them with young forests would reduce CO_2 and global warming.

What is the net effect of cutting old-growth forest on CO_2?

Although the carbon in wood cut from old-growth forests is placed in long-term storage as lumber in buildings, the decay of discarded leaves and branches and burning of resulting sawdust and wood chips produces a greater quantity of CO_2 than is lost. Thus, harvesting of old-growth forests contributes substantially to atmospheric CO_2.

Can replacement forests reduce the CO_2 gain from cutting of old forests?

Although the rapid growth of young trees removes CO_2, it may require up to two centuries to accumulate enough fixed carbon to equal that contributed by cutting of the old growth.

What other means of withdrawing CO_2 from the atmosphere is possible?

CO_2 may be removed from the atmosphere by planting trees on land areas barren of plants.

How is photosynthesis affected by rising CO_2 levels? Why is this important?

As atmospheric CO_2 concentration rises, photosynthesis is intensified, accelerating the growth of new forests and food crops. This presents an example of negative feedback—a process that helps maintain the status quo. That is, the higher the CO_2 level in the atmosphere, the greater is the rate at which plants remove it.

How large an area covered with trees would be required to remove the amount of new CO_2 now entering the air each year?

To remove the quantity of new CO_2 that now enters the air annually, some 7 million square kilometers of closed-canopy forest would be required. This is a land area about equal to the size of Australia.

How might forest regions in the U. S. change in response to global warming?

The loblolly pine range in the Southeast and the Douglas fir range in the Pacific Northwest might shift northward and to higher elevations.

SAVANNA BIOME

SAVANNA BIOME 534

Savanna Formation Classes

Savanna Formation Classes 534

Describe *savanna woodland.*

Savanna woodland: a formation class of the savanna biome in which trees are spaced rather widely apart, permitting development of a dense lower layer that usually consists of grasses. Landscape has open, parklike look.

With what climate is the savanna woodland closely associated?

Savanna woodland is closely associated with the wet-dry tropical climate (3).

Describe the trees of the savanna woodland.

Trees are of medium height, often show flattened crowns or umbrella shape, have rough, thick bark. Some species are small-leaved with thorns, others are broadleaf deciduous trees that shed leaves in cool dry season.

What is the importance of fire in the savanna woodland?

Fires of frequent occurrence in dry season consume grasses, but trees can survive. Fires and browsing animals may sustain grassland at expense of forest.

What is *thorntree-tall grass savanna?* Where does it occur?

Thorntree-tall grass savanna: a formation class transitional to the desert biome in Africa. More open than the savanna woodland. Common trees are acacia, baobab; typical grass is tall, coarse elephant grass.

With what climate is the thorntree-tall grass savanna identified?

Thorntree-tall grass savanna closely identified with semiarid subtype of dry tropical and subtropical climates (4s, 5s); also with semidesert subtypes (4sd, 5sd).

What soils are associated with the savanna biome?

Soils of savanna biome: Ustalfs, Ultisols, Oxisols, Vertisols. Ustalfs may get replenishment of bases from rain of fine dust from adjacent desert.

What is the *campo cerrado*?

Campo cerrado: in Brazilian Highlands, is a form of savanna with deep-rooted broadleaf evergreen trees.

Describe the savanna biome in eastern Australia.

Savanna biome in eastern Australia is *sclerophyllous tree savanna,* dominated by *eucalyptus trees*; soils are dominantly Vertisols.

DROUGHT AND LAND DEGRADATION IN THE AFRICAN SAHEL 535

What climatic hazard exists in the Sahelian zone?

Drought is a major environmental hazard in the African Sahelian zone (Sahel).

Define *drought.*

Drought: occurrence of substantially lower-than-average precipitation is a season that normally has ample precipitation for the support of food-producing plants.

In what way is the environment altered by intensive land use during a drought?

Overgrazing and trampling of land surface during a drought results in desertification, from which land surface may not recover in ensuing moist period or rainy season.

Define *desertification.*

Desertification: degradation of the quality of plant cover and soil as a result of overuse by humans and their domesticated animals.

GRASSLAND BIOME

GRASSLAND BIOME 536

What is *prairie?* What plants does it have?

Prairie: a formation class consisting of tall grasses as the dominant herbs, with forbs subdominant. Grasses are deeply rooted and dense. Forbs flower in late summer. In Midwest, dominant grasses are bluestems; typical forb is black-eyed susan.

With what climate is the tall-grass prairie identified?

Climate associations in North America: subhumid subtype of moist continental climate (10sh) and moist subtropical climate (6sh); extends into humid subtype of same climates.

What is the *Pampa*?

Pampa of Argentina and Uruguay is large region of tall-grass prairie.

What is the *Puszta*?

Puszta is tall-grass prairie region in Hungary.

With what soils is the tall-grass prairie identified?

Soils of tall-grass prairie largely Udolls, with some Ustolls in western portions.

Describe *steppe (short-grass prairie).*

Steppe (short-grass prairie): a formation class of short grasses occurring in sparsely distributed clumps or bunches; herbs also present. Scattered shrubs and low trees may also occur. Coverage is poor, much bare soil exposed. Plants of American steppe: buffalo grass, sunflower, loco weed.

With what regions and climate is steppe vegetation associated?

Major steppe grassland regions: North America and Eurasia, coinciding with semiarid subtype of dry midlatitude climate (9s). Small area in S. Africa (*veldt*).

With what soils is the steppe grassland associated?

Steppe grasslands associated with suborders of Mollisols: Borolls, Ustolls, Xerolls. Some areas of Aridisols in zone transitional to desert.

DESERT BIOME

DESERT BIOME 537

What is *thorntree semidesert?*

Thorntree semidesert: a formation class transitional between savanna and desert biomes in low latitudes, associated with tropical and subtropical climates.

Describe thorntree semidesert.

Thorntree semidesert consists of xerophytic, thorny trees and shrubs. Subtypes: thorn forest and thorn woodland; thorntree-desert grass savanna. Local names for such formations are *thorn forest, thornbush, thornwoods.*

With what climate is thorntree semidesert associated?

Climate association: semidesert and desert subtypes of dry tropical and dry subtropical climate (4sd, 4d, 5sd, 5d). Long dry season, very brief rainy season.

What is the *caatinga*? the *dornveldt*?

Caatinga is thorntree are of NE Brazil; *dornveldt* is thorntree area of S. Africa.

What soils dominate the regions of thorntree semidesert?

Dominant soils: Aridisols; some Ustalfs.

What is *semidesert?*

Semidesert: a formation class transitional between savanna biome (or short-grass prairie) and desert biome, ranging in latitude from tropical zone to midlatitude zone. Secondary formation subclasses: *Semidesert scrub and woodland (Dsd)*; and *semidesert scrub (Dss).*

Describe semidesert.

Semidesert is xerophytic scrub vegetation. Example: sagebrush semidesert of SW United States. Soils are largely Aridisols.

What is *dry desert?*

Dry desert: formation class of xerophytic plants widely dispersed and providing no important degree of ground cover.

On the world map name the two formation classes of dry desert. (See Figure 27.6 pg. 520.)

Two formation classes of dry desert on the world map include: *desert (D),* and in Australia, *desert alternating with porcupine grass semidesert (Dsp).*

Of what kinds of plants does dry desert consist?

Plants of dry desert are small, hard-leaved or spiny shrubs, succulent plants (cacti), or hard grasses. Also some ephemeral annuals.

Describe the desert plants of the Mojave-Sonoran desert.

Mojave-Sonoran desert region of SW United States features large treelike plants (saguaro cactus, pricklypear cactus and shrubs (ocotillo plant), creosote bush, smoke tree.

What plants are typical of the dry desert in Africa?

In Sahara Desert of Africa, typical plants are a hard grass (*Stipa*) and tamarisk (*Tamarix*) tree.

With what climate and soil is the dry desert associated?

Climate of dry desert is desert subtype of all three dry climates (4d, 5d, 9d). Soils are Aridisols, with large areas of Psamments and other suborders of Entisols.

TUNDRA BIOME

TUNDRA BIOME 538

What is *arctic tundra?*

Arctic tundra: formation class of the tundra biome occupying low areas of tundra climate (12) in lands fringing Arctic Ocean.

What is *alpine tundra?*

Alpine tundra: formation class of tundra found at high altitudes above timberline in mountains.

What conditions of soil and water drainage prevail in the tundra environment?

Tundra is underlain by permafrost, thaws in summer and is marshy. Frozen soil limits depth of root penetration, disrupts roots. Cold drying winds of winter abrade and reduce plants to low forms.

Name some plants of the arctic tundra.

Typical plants of arctic tundra: dwarf willow, sedges, mosses, lichens, flowering forbs.

With what soils is the arctic tundra associated?

Soils are various representatives of Inceptisols, Entisols, and Histosols. Large areas are Cryaquepts and Cryorthents. Histosols form in bog areas.

What is *fell-field?*

Fell-field: rocky pavement, usually formed into stone polygons and rings.

Describe the tundra ecosystem.

Tundra ecosystem has few species, but large numbers of individuals in each. Caribou are important grazers, also musk-oxen. Predators include wolves, wolverines, arctic fox, and polar bear. Smaller mammals are snowshoe rabbits, lemmings. Many insects and migratory birds.

ALTITUDE ZONES OF VEGETATION

ALTITUDE ZONES OF VEGETATION 539

How does increasing altitude influence the formation classes of vegetation present?

At increasing altitude, formation classes change to match climate of environment; succession of types resembles transect to high latitudes. High altitudes have needleleaf forest and alpine tundra.

CHAPTER 27—SAMPLE OBJECTIVE TEST QUESTIONS

A. MATCHING

1.	herbs	a. _______	air plants
2.	mosses	b. _______	rainforest
3.	epiphytes	c. _______	short grass
4.	graminoid	d. _______	dense shrub
5.	lianas	e. _______	forbs
6.	maquis	f. _______	grasslike
7.	steppe	g. _______	bryoids

B. MULTIPLE CHOICE

1. Which of the following is not one of the major biomes?

   ________a. forest.

   ________b. woodland.

   ________c. desert.

   ________d. tundra.

2. Soil orders most closely associated with the equatorial rainforest plant formation class are

   ________a. Udolls and Ustolls.

   ________b. Borolls and Xerolls.

   ________c. Oxisols and Ultisols.

   ________d. Spodosols and Histosols.

3. The taiga is essentially the same plant formation class as the

   ________a. arctic tundra.

   ________b. sclerophyll forest.

   ________c. thorntree semidesert.

   ________d. cold woodland.

C. COMPLETION

1. A biome consisting of a combination of trees and grassland is the ________________________.
2. Epiphytes that enclose a living tree and cause it to die are called ________________________.
3. A high-altitude rainforest closely akin to tropical rainforest is a ________________________ forest.

4. A tropical-zone forest with deciduous trees, found in a wet-dry climate, is a ______________ ______________ forest.
5. The common tree of sclerophyll forest in Australia is the ______________.
6. Besides tall grasses, the prairie formation class has flowering annuals called ______________ ______________.

ANSWERS TO SELF-TESTING QUESTIONS

CHAPTER 1

A. b–4, c–1, e–2, f–5, g–3.
B. 1–c, 2–a, 3–a, 4–d.
C. 1. gravitation, 2. meridian, 3. great circle, 4. nautical mile, 5. mean solar, 6. revolution.

CHAPTER 2

A. a–8, b–4, c-6, d–9, e-7, f-1, g-2, h–10, i-3, j–5.
B. 1–b, 2–c, 3–d, 4–d, 5–c.
C. 1. meteorology, 2. nitrogen; 78%, 3. oxygen; 21%, 4. carbon dioxide, 5. stratosphere; troposphere, 6. ozone; ultraviolet, 7. magnetosphere; solar wind, 8. sensible heat.

CHAPTER 3

A. a–8, b–4, c–10, d–7, e-3, f–9, g–5, h–2, i–6, j–1.
B. 1–b, 2–c, 3–a.
C. 1. radiation balance, 2. insolation, 3. equatorial; subtropical; subarctic, 4. albedo, 5. counterradiation.

CHAPTER 4

A. a–4, b–1, c–5, d–2, e–3.
B. 1–c. 2–a 3–d.
C. 1. epilimnion, 2. killing frost, 3. annual range, 4. isotherm, 5. ice sheets, 6. carbon dioxide.

CHAPTER 5

A. a–8, b–10, c–4, d–6, e–9, f–1, g–3, h–2, i–7, j–5.
B. 1–b, 2–c, 3–a.
C. 1. isobaric surface, 2. sea breeze, 3. Hadley cell, 4. jet stream, 5. west-wind drift.

CHAPTER 6

A. a–7, b–8, c–1, d–9, e–2, f–10, g–3, h–6, i–5, j–4.
B. 1–b, 2–a, 3–c.
C. 1. specific humidity, 2. dry adiabatic, 3. radiation, 4. chinook, 5. smog, 6. washout, 7. photochemical.

CHAPTER 7

A. a–3, b–10, c–6, d–8, e–2, f–9, g–4, h–1, i–7, j–5.
B. 1–a, 2–c, 3–d.
C. 1. source regions, 2. warm, 3. cyclone family, 4. polar outbreak, 5. storm surge.

CHAPTER 8

A. 1–d, 2–b, 3–d, 4–a, 5–b.
B. 1. continentality, 2. midlatitude, 3. climograph, 4. steppe.

CHAPTER 9

A. a–3, b–5, c–2, d–1, e–4.
B. 1–d, 2–b, 3–a.
C. 1. subsurface, 2. evapotranspiration, 3. intermediate, 4. storage withdrawal, 5. water deficit, 6. soil-water shortage.

CHAPTER 10

A. 1–d, 2–b, 3–d, 4–b,
B. 1. rain green 2. bush-fallow farming, 3. broadleaf-evergreen, 4. thorntree semidesert.

CHAPTER 11

A. 1–c, 2–a, 3–c, 4–b.
B. 1. steppe, 2. Mollisols, 3. peat.

CHAPTER 12

A. a–9, b–7, c–1, d–10, e–4, f–8, g–2, h–3, i–6, j–5.
B. 1–c, 2–d, 3–a.
C. 1. olivine, 2. gabbro (basalt), 3. rhyolite, 4. hydrolysis, 5. clastic.

CHAPTER 13

A. a–7, b–5, c–6, d–1, e–3, f–8, g–4, h–2.
B. 1–b, 2–a, 4–c.
C. 1. core, mantle, crust; 2. asthenosphere, 3. subduction, 4. transform, 5. six; Pacific, 6. axial rift.

CHAPTER 14

A. a–1, b–3, c–2, d–4, e–5.
B. 1–d, 2–a, 3–c.
C. 1. composite, 2. shield, 3. cinder cone, 4. transcurrent.

CHAPTER 15

A. a–7, b–4, c–10, d–8, e–1, f–9, g–2, h–5, i–3, j–6.
B. 1–a, 2–b, 3–a.
C. 1. talus slope, 2. exfoliation dome, 3. saprolite, 4. solifluction, 5. scarification.

CHAPTER 16

A. a–7, b–9, c–5, d–2, e–1, f–8, g–4, h–3, i–10, j–6.
B. 1–c, 2–b, 3–b.
C. 1. floodplain, 2. eutrophication, 3. sulfuric, 4. chloride, 5. hydrograph, 6. base flow.

CHAPTER 17

A. a–8, b–6, c–4, d–1, e–7, f–2, g–3, h–5.
B. 1–b, 2–a, 3–d.
C. 1. infiltration, 2. sediment yield, 3. hydraulic; bank caving; abrasion; corrosion, 4. capacity, 5. base level.

CHAPTER 18

A. a–3, b–4, c–6, d–1, e–7, f–5, g–2.
B. 1–b, 2–c, 3–d.
C. 1. peneplain, 2. isostasy, 3. rejuvenation, 4. playa, 5. pediplain.

CHAPTER 19

A. a–6, b–8, c–9, d–4, e–7, f–3, g–2, h–5, i–1.
B. 1–d, 2–a, 3–c.
C. 1. dip, 2. hogbacks, 3. dendritic, 4. antecedent.

CHAPTER 20

A. a–7, b–4, c–8, d–3, e–1, f–5, g–9, h–10, i–6, j–2.
B. 1–b, 2–c, 3–d.
C. 1. raised shoreline, 2. retrogradation, 3. peat, 4. ebb, 5. fault, 6. delta, 7. tidal inlet, 8. swash.

CHAPTER 21

A. a–8, b–9, c–5, d–2, e–10, f–4, g–3, h–7, i–6, j–1.
B. 1–c, 2–b, 3–b.
C. 1. sand sea, 2. blowout, 3. foredunes, 4. loess, 5. Dust Bowl.

CHAPTER 22

A. a–9, b–4, c–8, d–10, e–1, f–3, g–7, h–2, i–5, j–6.
B. 1–c, 2–d, 3–a.
C. 1. abrasion; plucking, 2. horn, 3. fiord, 4. Wisconsinan, 5. stratified drift; till, 6. interlobate moraine.

CHAPTER 23

A. a–3, b–5, c–6, d–1, e–4, f–7, g–2.
B. 1–c, 2–b, 3–d.
C. 1. polypedon; pedons, 2. humus; humification, 3. layered, 4. base.

CHAPTER 24

A. a–5, b–3, c–6, d–1, e–7, f–2, g–4.
B. 1–c, 2–c, 3–a.
C. 1. epipedon; plaggen, 2. natric; salic, 3. sand, 4. diagnostic horizons.

CHAPTER 25

A. a–9, b–1, c–6, d–2, e–7, f–8, g–3, h–4, i–5.
B. 1–c, 2–d, 3–a.
C. 1. respiration; energy, 2. net photosynthesis; primary productivity, 3. 10%; 50%, 4. macronutrients; hydrogen; carbon; and oxygen.

CHAPTER 26

A. a–3, b–5, c–6, d–7, e–1, f–4, g–2.
B. 1–a, 2–c, 3–d.
C. 1. xerophytes, 2. guard; transpiration, 3. ephemeral, 4. hibernation, 5. photoperiod, 6. bioclimatic.

CHAPTER 27

A. a–3, b–5, c–7, d–6, e–1, f–4, g–2.
B. 1–b, 2–c, 3–d.
C. 1. savanna, 2. stranglers, 3. montane, 4. monsoon, 5. eucalyptus, 6. forbs.